LONDON MATHEMATICAL SOCIETY LECTURE NOTE SERIES

Managing Editor: PROFESSOR I.M. JAMES
24-29 St Giles, Oxford

Prospective authors should contact the

Already published in this series

1. General cohomology theory and K-theory, PETER HILTON.
4. Algebraic topology: A student's guide, J.F. ADAMS.
5. Commutative algebra, J.T. KNIGHT.
7. Introduction to combinatory logic, J.R. HINDLEY, B. LERCHER and J.P. SELDIN.
8. Integration and harmonic analysis on compact groups, R.E. EDWARDS.
9. Elliptic functions and elliptic curves, PATRICK DU VAL.
10. Numerical ranges II, F.F. BONSALL and J. DUNCAN.
11. New developments in topology, G. SEGAL (ed.).
12. Symposium on complex analysis, Canterbury, 1973, J. CLUNIE and W.K. HAYMAN (eds.).
13. Combinatorics, Proceedings of the British combinatorial conference 1973, T.P. McDONOUGH and V.C. MAVRON (eds.).
14. Analytic theory of abelian varieties, H.P.F. SWINNERTON-DYER.
15. An introduction to topological groups, P.J. HIGGINS.
16. Topics in finite groups, TERENCE M. GAGEN.
17. Differentiable germs and catastrophes, THEODOR BROCKER and L. LANDER.
18. A geometric approach to homology theory, S. BUONCRISTIANO, C.P. ROURKE and B.J. SANDERSON.
19. Graph theory, coding theory and block designs, P.J. CAMERON and J.H. VAN LINT.
20. Sheaf theory, B.R. TENNISON.
21. Automatic continuity of linear operators, ALLAN M. SINCLAIR.
22. Presentations of groups, D.L. JOHNSON.
23. Parallelisms of complete designs, PETER J. CAMERON.
24. The topology of Stiefel manifolds, I.M. JAMES.
25. Lie groups and compact groups, J.F. PRICE.
26. Transformation groups: Proceedings of the conference in the University of Newcastle upon Tyne, August 1976, CZES KOSNIOWSKI.
27. Skew field constructions, P.M. COHN.
28. Brownian motion, Hardy spaces and bounded means oscillation, K.E. PETERSEN.
29. Pontryagin duality and the structure of locally compact abelian groups, SIDNEY A. MORRIS.
30. Interaction models, N.L. BIGGS.
31. Continuous crossed products and type III von Neumann algebras, A. VAN DAELE.
32. Uniform algebras and Jensen measures, T.W. GAMELIN.
33. Permutation groups and combinatorial structures, N.L. BIGGS and A.T. WHITE
34. Representation theory of Lie groups, M.F. ATIYAH.
35. Trace ideals and their applications, BARRY SIMON.

London Mathematical Society Lecture Note Series. 42

Topics in the
Theory of Group Presentations

D. L. Johnson

CAMBRIDGE UNIVERSITY PRESS

CAMBRIDGE

LONDON   NEW YORK   NEW ROCHELLE

MELBOURNE   SYDNEY

CAMBRIDGE UNIVERSITY PRESS
Cambridge, New York, Melbourne, Madrid, Cape Town, Singapore, São Paulo

Cambridge University Press
The Edinburgh Building, Cambridge CB2 8RU, UK

Published in the United States of America by Cambridge University Press, New York

www.cambridge.org
Information on this title: www.cambridge.org/9780521231084

First published 1980
Re-issued in this digitally printed version 2008

A catalogue record for this publication is available from the British Library

ISBN 978-0-521-23108-4 paperback

To Eileen

Contents

Preface

These notes arose from a course of lectures given to final-year and postgraduate students at the University of Nottingham, and comprise a substantially revised and extended version of my earlier contribution to this series. As before, the emphasis is on concrete examples of groups exhibited in their natural settings and thus to demonstrate at a modest level some of the pervasive connections between group theory and other branches of mathematics. Such is the current rate of progress (both upwards and outwards) in combinatorial group theory that no attempt at completeness is feasible, though it is hoped to bring the reader to within hailing distance of the frontiers of research in one ot two places.

My thanks are due to a host of colleagues, students and friends whose names, too numerous to mention here, may be found scattered through the ensuing pages. It is a pleasure to acknowledge a special debt of gratitude to Professor Sandy Green for introducing me to research mathematics, to Dr E.F. Robertson for his encouragement and for help in correcting the proofs, and to Dr H.R. Morton for much valuable advice on the final chapter. My thanks also go to Professor I.M. James for keeping a paternal eye on things, to Mrs Anne Towndrow for typing half the manuscript, and to the staff of Cambridge University Press for their speed and skill in setting the text (especially the other half).

1· Free groups and free presentations

> The words are all there ready; now we've got
> to get them in the right order. (Python)

A group G is generated by a subset X if each of its elements
can be expressed as a product of members of $X^{\pm 1}$. Such a product
is called a word, and a relation is an equation between two words.
A set R of relations that hold in G defines the group if every
relation that holds in G is a consequence of R . When this
happens, we say that G is presented by X and R . This definition is made rigorous using the concept of a free group (essentially, a group having a set of generators between which there are
no non-trivial relations), which is defined using a universal
property. Having developed some elementary but important properties of free groups (such as their existence), we proceed to the
fundamental theorem of §2, where Schreier's proof is given in detail and Nielsen's original method in outline. In §3, the definition of group presentation is made rigorous, and this is used
to clarify the proof of the Nielsen-Schreier theorem by means of
an anotated example. §4 explains how to pass from a group multiplication table to a presentation and from one presentation to
another, as well as describing a presentation for a direct product
of two groups.

§1. Elementary properties of free groups

The fundamental notion used in defining presentations of groups
is that of a free group. As the definition suggests, the idea of
freeness is applicable in algebraic situations other than group
theory.

Definition 1. A group F is said to be *free* on a subset $X \subseteq F$
if, for any group G and any mapping $\theta : X \to G$, there is a

1

unique homomorphism $\theta' : F \to G$ such that

$$x\theta' = x\theta \tag{1}$$

for all $x \in X$. The cardinality of X is called the *rank* of F .

Remark 1.   There are various ways of expressing the property (1).
We may say that θ' extends θ or that θ' agrees with θ on
X or, letting $\iota : X \to F$ denote inclusion, that the following
diagram is commutative:

In general, a diagram involving sets and mappings is called commutative if any two composite mappings, beginning at the same
place and ending at the same place in the diagram, are equal.   In
this case, this boils down to the single assertion that $\iota\theta' = \theta$.

Remark 2.   There is an analogy between this situation and a familiar one encountered in linear algebra; let V be a vector space
over a field k and B a basis for V .   Then for any vector space
W over k and any mapping $\tau : B \to W$, there is a unique k-linear
transformation $\tau' : V \to W$ extending τ .   This property is known as
'extension by linearity' and can be used to *define* the notion of
basis.

Remark 3.   If we write 'abelian group' in place of 'group' in the
two places where this word appears in Definition 1, we obtain the
definition of a free abelian group.   A free abelian group of rank
ω is just the direct sum of ω infinite cyclic groups (proved
for finite ω in Theorem 6.2).

Remark 4.   By convention, we take E (the trivial group) to be
free of rank 0 , the subset X being empty.   The infinite cyclic
group $\{x^n | n \in Z\}$ is free of rank 1 .   We denote it by Z as it

is just the multiplicative version of the additive group of integers. Take $X = \{x\}$, and given

$$\left.\begin{array}{l} \theta \,:\, X \to G \\ x \mapsto y \end{array}\right\} \,,$$

simply define for all $n \in Z$,

$$x^n \theta' \;=\; y^n \;.$$

θ' is obviously a homomorphism extending θ , while if θ'' is another,

$$x^n \theta'' = (x\theta'')^n = y^n = (x\theta')^n = x^n \theta' \;,$$

proving that θ' is unique.

Remark 5. There are one or two things to check before this definition can have any value. One can show for example that there does exist a free group of any given rank, and that the rank of a free group is well defined. These together with other elementary properties of free groups form the content of our first four theorems.

Theorem 1. (i) *If* F *is free on* X *, then* X *generates* F *.*
(ii) *Two free groups of the same rank are isomorphic.*
(iii) *Free groups of different ranks are not isomorphic.*

Proof. (i) Recall that if X is a subset of a group G , the intersection of all subgroups of G containing X is called the subgroup generated by X and written $<X>$. We leave it as an exercise to show that this coincides with the set of all finite products of members of X and their inverses. Returning to the case in hand, we let $<X>$ play the role of G in Definition 1, taking θ to be inclusion. Letting ϕ denote the inclusion of $<X>$ in F , we have the following picture:

Since this diagram commutes, we have $\iota\theta'\phi = \theta\phi = \iota$, so that
$\theta'\phi : F \to F$ extends ι . But so does 1_F , and so by the uniqueness part of Definition 1 (with ι, F in place of θ, G), we have
$\theta'\phi = 1_F$, whence ϕ is onto and $\langle X \rangle = F$, as required.

    (ii) Let F_j be free on X_j and let $\iota_j : X_j \to F_j$ denote in-
clusion, $j = 1,2$. Assume that $|X_1| = |X_2|$, so that there is a
bijection $\kappa : X_1 \to X_2$. Let α, β be the homomorphisms extending
$\kappa\iota_2, \kappa^{-1}\iota_1$ as in the following diagrams:

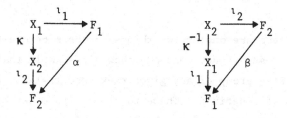

Now $\iota_1\alpha\beta = \kappa\iota_2\beta = \kappa\kappa^{-1}\iota_1 = \iota_1$, so that $\alpha\beta : F_1 \to F_1$ extends
ι_1 . But 1_{F_1} also extends ι_1 , so uniqueness implies $\alpha\beta = 1_{F_1}$.
Similarly, $\beta\alpha = 1_{F_2}$, and α is the required isomorphism.

    (iii) Let F be free on a subset X with $|X| = \omega$, and let
G be any group. Then it is the burden of Definition 1 that the
mappings: $X \to G$ are in one-to-one correspondence with the homomorphisms: $F \to G$. Thus, there are exactly 2^ω homomorphisms
from F to Z_2 . Since this number is invariant under isomorphism, we see that 2^ω , and hence the rank ω , is determined by
the isomorphism class of F .

Theorem 2. *There exists a free group of any given rank.*

4

Proof.   We construct the 'group of words' $F = F(X)$ on a given
set X , and prove that it is free of rank |X| . The free group F
on a given set X is constructed as follows. Let $\hat{X} = \{\hat{x} | x \in X\}$
be any set in one-to-one correspondence with, and disjoint from,
X and put $T = X \cup \hat{X}$. If T^n denotes the nth Cartesian power
of T (n = 0,1,2,...), put $W = \bigcup_{n \geq 0} T^n$, the set of words in X .
A word $w \in T^n$ is said to have length n , and the single element
of T^0 is called the empty word and denoted by e . A word
$w = (t_1, \ldots, t_n)$ in W is called reduced if there is no i between 1 and n-1 such that $\hat{t}_i = t_{i+1}$, interpreting $\hat{\hat{s}} = s$.
Letting F be the set of reduced words, it is clear that $e \in F$
and $X \subseteq F$. The product of two reduced words of positive length

$$a = (x_1, \ldots, x_m), \; b = (y_1, \ldots, y_n)$$

is defined to be

$$(x_1, \ldots, x_{m-k}, \; y_{k+1}, \ldots, y_n) \quad ,$$

where k is the largest integer such that none of the words

$$(x_m, y_1), \ldots, (x_{m-r+1}, y_r)$$

are reduced, while we = ew = w for any word w . It is clear
that this defines a binary operation on F for which e is an
identity and $(x_1, \ldots, x_m)^{-1} = (\hat{x}_m, \ldots, \hat{x}_1)$. The tricky bit, surprisingly enough, is the proof of the associative law. Now take
three words in F :

$$a = (x_1, \ldots, x_\ell), \; b = (y_1, \ldots, y_m), \; c = (z_1, \ldots, z_n) \quad .$$

If any of ℓ, m, n are zero, we clearly have (ab)c = a(bc) , so
assume they are all positive. Supposing that the lengths of ab
and bc are $\ell + m - 2r$ and $m + n - 2s$ respectively, we distinguish
three cases. First, if r+s < m , both (ab)c and a(bc) are

equal to the reduced word

$$(x_1, \ldots, x_{\ell-r}, y_{r+1}, \ldots, y_{m-s}, z_{s+1}, \ldots, z_n) \quad ;$$

secondly, if $r+s = m$, both are equal to $\alpha\epsilon$, where

$$\alpha = (x_1, \ldots, x_{\ell-r}), \quad \epsilon = (z_{s+1}, \ldots, z_n) \quad .$$

Finally, in the case $r+s > m$, we define

$$\beta = (x_{\ell-r+1}, \ldots, x_{\ell-m+s}) = (y_{m-s+1}, \ldots, y_r)^{-1} = (z_{m-r+1}, \ldots, z_s) \ ,$$

$$\gamma = (x_{\ell-m+s+1}, \ldots, x_\ell) = (y_1, \ldots, y_{m-s})^{-1} \ ,$$

$$\delta = (z_1, \ldots, z_{m-r}) = (y_{r+1}, \ldots, y_m)^{-1} \ .$$

Thus, $a = \alpha\beta\gamma$, $b = \gamma^{-1}\beta^{-1}\delta^{-1}$, $c = \delta\beta\epsilon$, since the brackets can
safely be ignored by the first case handled above. Now by the
rule for forming products,

$$(ab)c = (\alpha\delta^{-1})(\delta\beta\epsilon) = \alpha(\beta\epsilon) \ , \quad \text{and}$$

$$a(bc) = (\alpha\beta\gamma)(\gamma^{-1}\epsilon) = (\alpha\beta)\epsilon \ ,$$

and again by the first case, these both coincide with the reduced
word $(x_1, \ldots, x_{\ell-m+s}, z_{s+1}, \ldots, z_n)$.

We now simplify the notation by dropping the commas and brackets
and writing x^{-1} for $\hat{x}$ $(x \in X \cup \hat{X})$, so that if ι is the inclusion of X in F , all we have to do is check Definition 1 ver-
batim. If G is a group and $\theta : X \to G$ a mapping, define

$$e\theta' = e \ , \quad x^{-1}\theta' = (x\theta)^{-1} \ ,$$

$$(x_1 \ldots x_n)\theta' = x_1\theta' \ldots x_n\theta' \ ,$$

for any $x \in X$ and any reduced word $x_1 \ldots x_n$. It is a routine matter to check that θ' is a homomorphism extending θ . If θ'' is another, it must agree with θ' on X and since X plainly generates F , we must have $\theta' = \theta''$.

Theorem 3. *Let* F *be a group and* X *a subset of* F *; then* F *is free on* X *if and only if the following two conditions hold:*

   (i) X *generates* F *,*

   (ii) there is no non-trivial relation between the elements of X *, that is, if for* $n \in N$ *,* $x = x_1 \ldots x_n$ *where for all* i *, either* $x_i \in X$ *or* $x_i^{-1} \in X$ *, and for all* i *with* $1 \le i \le n-1$ *,* $x_i x_{i+1} \ne e$ *, then* $x \ne e$ *.*

Proof. First suppose that F is free on X , so that X generates F by Theorem 1(i). Now let $X' = \{x' | x \in X\}$ be an abstract copy of X and consider the group of words $F(X')$ as constructed in the proof of Theorem 2. By Definition 1, the priming map : $X \to F(X')$ extends to a homomorphism : $F \to F(X')$ under which any reduced word $x \in F$ is mapped to a reduced word in $F(X')$ of the same length. Thus no reduced word in F of length $n \le 1$ can be e , since this certainly holds in $F(X')$.

   For the converse, note that conditions (i) and (ii) imply that every member of F is uniquely expressible as a reduced word in $X \cup X^{-1}$. The freeness of F on X is now verified in just the same way as that of $F(X)$ on X in the final part of the proof of Theorem 2.

Theorem 4. *If* X *is a set of generators for a group* G *and* $F(X)$ *is the group of words in* X *, then there is an epimorphism* $\theta: F(X) \to G$ *fixing* X *elementwise. Every group is a homomorphic image of some free group.*

Proof. The required epimorphism is just the (unique) extension to the free group $F(X)$ of the inclusion : $X \to G$; it is onto because $X \subseteq \text{Im } \theta \le G$ and $\langle X \rangle = G$. The second assertion now follows from the simple observation that any group G *has* a set of generators, for example, $G = \langle G \rangle$.

EXERCISE 1. Let X be a subset of a group G . Prove that $\langle X \rangle$ is equal to the set of all finite products of members of X and their inverses. Deduce that if two homomorphisms from G to a group H agree on a set X of generators of G (i.e. $\langle X \rangle = G$), then they are equal.

EXERCISE 2. Given groups G and H , a subset X of G and a homomorphism $\theta: G \to H$, prove that $\langle X\theta \rangle = \langle X \rangle \theta$. Defining

$$d(G) = \min\{|X| \mid X \subseteq G, \langle X \rangle = G\} \quad,$$

prove that for any homomorphic image H of G , $d(H) \leq d(G)$.

EXERCISE 3. Given a subset X of a group G , define the normal closure $\bar{X}$ of X to be the intersection of all *normal* subgroups of G containing X . Prove that $\bar{X}$ is just the set of all finite products of conjugates of members of X and their inverses. If H is a group and $\theta: G \to H$ an epimorphism, show that $\overline{X\theta} = \bar{X}\theta$.

EXERCISE 4. Let F be a free group of rank ω and G a group isomorphic to F . Prove that G is free of rank ω .

EXERCISE 5. A group G has a normal subgroup N such that G/N is free. Prove that G has a subgroup F such that FN = G and $F_\cap N = E$. (Such an F is called a *complement* for N in G .)

EXERCISE 6. Call a group P *projective* if given any epimorphism $\nu: B \to C$ of groups and any homomorphism $\phi: P \to C$, there is a homomorphism $\mu: P \to B$ such that $\phi = \mu\nu$:

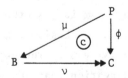

8

Prove that P is projective if and only if P is free.

EXERCISE 7. Call a group I *injective* if given any monomorphism
$\iota : A \to B$ of groups and any homomorphism $\phi : A \to I$, there is a
homomorphism $\mu : B \to I$ such that $\phi = \iota\mu$:

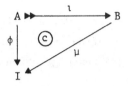

Prove that I is injective if and only if I is trivial. (Hint
(D.B.A. Epstein): Assume that the free group $A = F(x,y)$ is a
subgroup of a group $B = \langle x,y,z \rangle$ in which $z^{-1}xz = y$, $z^{-1}yz = x$,
$z^2 = e$.)

§2. The Nielsen-Schreier theorem

The first step in proving the Basis Theorem for finitely-
generated abelian groups (see §6 below) is to show, at least in
the case of finite rank, that subgroups of free abelian groups are
free abelian, and that the rank of the subgroup does not exceed
the rank of the group. This is a classical result of Dedekind,
and our purpose here is to prove its non-abelian analogue, taking
care to point out that the assertion about ranks does *not* hold in
the non-abelian case. We shall consider the free group $F = \Gamma(X)$
on an arbitrary set X , invoking the Axiom of Choice to assert
that X can be well-ordered. The intuitionistic reader is free
to assume that X is finite, since this is the only case of
practical interest to us.

The proof we give is essentially due to Schreier, and is
divided into a number of steps, the most important for the sequel
(§12) being embodied in Lemma 3.

1. The ordering of F . Given that X is well-ordered, so
is X^{-1} and so also is $T = X \cup X^{-1} = X^{\pm 1}$; for example, if
$x,y \in X$, define $x < y^{-1}$, and

9

$$x^{-1} < y^{-1} \iff x < y \quad .$$

Now the elements of F are words of the form

$$w = x_1 \ldots x_n \ , \ x_i \in T \ , \ x_i x_{i+1} \neq e \ ;$$

we call n the *length* of the word w , $n = \ell(w)$, and take $\ell(e) = 0$. For $v, w \in F$, we define $v < w$ if $\ell(v) < \ell(w)$, and order words of equal length lexicographically, that is, if

$$v = x_1 \ldots x_n \neq w = y_1 \ldots y_n \ , \quad x_i, y_i \in T \ ,$$

and m is least such that $x_m \neq y_m$, we define

$$v < w \iff x_m < y_m \quad .$$

The result is easily seen to be a well-ordering of F . For example, if $X = \{x, y\}$, then with respect to the ordering $x < y < x^{-1} < y^{-1}$ of T , the first few elements of F are $e < x < y < x^{-1} < y^{-1} < x^2 < xy < xy^{-1} < yx < y^2 < yx^{-1} < x^{-1}y < x^{-2} < x^{-1}y^{-1} < y^{-1}x < y^{-1}x^{-1} < y^{-2} < x^3 < x^2y < x^2y^{-1} < xyx$. Note that in any subgroup H of F , the least element is always e (since this is least in F) and if, for example, $H = \langle xyx \rangle$, the least element of the coset Hxy^2 is $x^{-1}y$. That F is well-ordered is particularly easy to see in the case of finite rank, for then there are only finitely many words of any given length.

Lemma 1. *Let*

$$w = x_1 \ldots x_n, \ x_i \in T, \ n \geq 1 \ ,$$

be a reduced word in F ; *then, for* $v \in F$,

$$v < x_1 \ldots x_{n-1} \Rightarrow vx_n < w \quad .$$

10

Proof. If $\ell(v) < n-1$, then $\ell(vx_n) \leq \ell(v) + 1 < n = \ell(w)$, and
the result holds. Otherwise

$$v = y_1 \cdots y_{n-1}, \ y_i \in T \ ,$$

and there is a least m such that $y_m \neq x_m$ and then $y_m < x_m$.
If $y_{n-1}x_n = e$, $\ell(v) = n-2 < n = \ell(w)$ and we are done. If not

$$vx_n = y_1 \cdots y_{n-1}x_n \ ,$$

and, since $\ell(vx_n) = \ell(w)$, $y_1 = x_1, \ldots, y_{m-1} = x_{m-1}$ and
$y_m < x_m$, we have $vx_n < w$, as required.

2. The Schreier transversal. We fix a subgroup H of F
once and for all. Recall that a right coset of H in F is a
subset of F of the form

$$Hx = \{hx | h \in H\}$$

for $x \in F$. The key property of cosets is that

$$Hx = Hy \ \text{or} \ Hx \cap Hy = \emptyset, \ x,y \in F \ .$$

The cosets of H thus yield a partition of F , and we can find
a subset U of F such that, for any $x \in F$, there is exactly
one element $u \in U$ for which $x \in Hu$, that is,

$$F = \mathop{\cup}_{u \in U} Hu \ .$$

Such a subset U is called a (right) transversal for H in F .
The Schreier transversal (with respect to the given ordering of
T) is obtained if the representative in U of each coset is
taken to be the least element of that coset. Alternatively, we
list the cosets Hx as x runs over F in ascending order, thus:

$$He, Hx_1, Hx_2, \ldots$$

and, for each $x \in F$, delete Hx from the list if $Hx = Hy$ for some $y < x$. We then put U = the set of x such that Hx remains in the list. By construction then, the transversal U has the property:

$$x < y, \; Hx = Hy \implies y \notin U \;, \tag{1}$$

from which follows the so-called Schreier property, which we now derive.

Lemma 2 (Schreier property). *Let* $x_1 \ldots x_n$ *be a reduced word in* F $(n \geq 1)$; *then*

$$x_1 \ldots x_n \in U \implies x_1 \ldots x_{n-1} \in U \;.$$

Proof. To prove the contraposed assertion, suppose that $x_1 \ldots x_{n-1} \notin U$. It follows from the fact that U is a transversal that there is a $u \in U$ such that

$$Hu = Hx_1 \ldots x_{n-1} \;,$$

whence, by definition of U, $u < x_1 \ldots x_{n-1}$. Thus by Lemma 1, $ux_n < x_1 \ldots x_n$. Now let $v \in U$ be such that $Hv = Hux_n$, so that $v \leq ux_n$. Thus, $v < x_1 \ldots x_n$ and $Hv = Hx_1 \ldots x_n$, so (1) implies that $x_1 \ldots x_n \notin U$, as required.

We call any transversal with the property of Lemma 2 a Schreier transversal, and note that any such transversal contains the identity e . While for the sake of definiteness we continue to let U denote the transversal consisting of the least element in each coset, we make the following important observation: the proof of Lemma 3 below requires only that U is a transversal containing e , while that of Lemma 4 depends only on the fact that U is a transversal with the Schreier property. Thus, *any* Schreier transversal leads to subsets A and B (constructed below) with the required properties.

3. The subset A of H. Let $u \in U$, $x \in T$; then as $ux \in F$ and U is a transversal for H in F, there is just one $v \in U$ such that $ux \in Hv$. Since v depends on u and x, we denote it by $\overline{ux}$. Since $ux = hv$ for some $h \in H$, we have $ux\overline{ux}^{-1} = h \in H$, for all $u \in U$, $x \in T$. We put

$$A = \{ux\overline{ux}^{-1} \mid u \in U, x \in T\} \ ,$$

which is a subset of H.

Lemma 3.   *The set A just defined generates H.*

Proof.   Let $x \in H$; since $x \in F$, we can write it as

$$x_1 \ldots x_n, \ x_i \in T \ ,$$

a reduced word. We define a sequence of elements $u_1, \ldots, u_{n+1} \in U$ inductively as follows:

$$u_1 = e, \ u_{i+1} = \overline{u_i x_i}, \ i \geq 1 \ .$$

Consider the sequence:

$$a_i = u_i x_i u_{i+1}^{-1}, \ 1 \leq i \leq n \ .$$

By definition, each $a_i \in A$ and so H contains

$$a_1 \ldots a_n = u_1 x_1 x_2 \ldots x_n u_{n+1}^{-1} = x u_{n+1}^{-1} \ .$$

Since $x u_{n+1}^{-1}$ and x both belong to H, so does u_{n+1}, and also $u_{n+1} \in U$; thus $u_{n+1} = e$, since e is the (unique) element of U representing the trivial coset H. Hence, $x = a_1 \ldots a_n$, proving that A generates H, as claimed.

4. Further properties of A. These are compressed into a lemma.

Lemma 4. *If* $u \in U$ *and* $x \in T = X^{\pm 1}$, *then*

(i) $ux\overline{ux}^{-1} = e$ *if and only if* $ux \in U$

(ii) $u = \overline{uxx^{-1}}$

(iii) *let* $ux, vy \in UT \setminus U$; *then either:* (a) $x\overline{ux}^{-1}vy = e$, *in which case* $v = \overline{ux}$, $y = x^{-1}$, $u = \overline{vy}$, *or* (b) *the reduced word in* F *representing* $w = x\overline{ux}^{-1}vy$ *has length at least two, begins with* x *and ends with* y ;

(iv) *the words* $ux\overline{ux}^{-1}$, $ux \notin U$, *are all distinct and the set of them is equal to* $B \stackrel{.}{\cup} B^{-1}$, *where* $B = \{ux\overline{ux}^{-1} \mid u \in U, x \in X\} \setminus \{e\}$.

Proof.

(i) $ux\overline{ux}^{-1} = e \iff ux = \overline{ux} \iff ux \in U$.

(ii) As $Hux = H\overline{ux}$, we have

$$Hu = H\overline{ux}x^{-1} = H\overline{uxx}^{-1} \quad ,$$

and so $u = \overline{uxx}^{-1}$, since both lie in U .

(iii) This assertion is the crux of the whole proof and it depends heavily but solely on the Schreier property. Let

$$\overline{ux} = r_1 \ldots r_m \quad , \quad v = t_1 \ldots t_n$$

be reduced words (all $r_i, t_j \in T$), so that

$$w = xr_m^{-1} \ldots r_1^{-1} t_1 \ldots t_n y \quad .$$

We examine this word minutely;

$$xr_m^{-1} = e \Rightarrow \overline{ux}x^{-1} = r_1 \ldots r_{m-1} \in U \quad , \text{ by Lemma 2,}$$

$$\Rightarrow u = \overline{\overline{ux}x}^{-1} \quad , \text{ by part (ii),}$$

$$= \overline{uxx}^{-1}$$

$$\Rightarrow ux = \overline{ux}$$

14

$\Rightarrow ux \in U$, by part (i),

contrary to assumption. Similarly, if $t_n y = e$, then
$vy = t_1 \ldots t_{n-1} \in U$, by Lemma 2, which again is not allowed.
  Thus we see that xr_m^{-1} and $t_n y$ are both reduced words. Let
$\overline{ux}^{-1}v$ be equal to the reduced word

$$ r_m^{-1} \ldots r_{i+1}^{-1} t_{i+1} \ldots t_n \quad . $$

There are four cases to consider:
(1) $i < m$ and $i < n$: then case (b) holds.
(2) $i = m < n$: here w is the reduced word

$$
\begin{cases}
xt_{m+1} \ldots t_n y \text{ , and (b) holds, or} \\[2mm]
t_{m+2} \ldots t_n y \text{ , whereupon:}
\end{cases}
$$

$\quad xt_{m+1} = e \Rightarrow \overline{ux}x^{-1} = t_1 \ldots t_{m+1} \in U$, by Lemma 2

$\qquad\qquad \Rightarrow \overline{ux}x^{-1} = u$, by parts (ii) and (i),

$\qquad\qquad \Rightarrow ux = \overline{ux}$

$\qquad\qquad \Rightarrow ux \in U$, which is not allowed.

(3) $i = n < m$: here w is the reduced word

$$
\begin{cases}
xr_m^{-1} \ldots r_{n+1}^{-1} y \text{ , and (b) holds, or} \\[2mm]
xr_m^{-1} \ldots r_{n+2}^{-1} \text{ , whereupon:}
\end{cases}
$$

$\quad r_{n+1}^{-1} y = e \Rightarrow vy = r_1 \ldots r_{n+1} \in U$, by Lemma 2,

contradicting our hypotheses.

(4) $i = m = n$: here w is equal to either xy (and (b) holds),
or to e , whereupon $y = x^{-1}$, $\overline{ux} = v$, and so by part (ii),

$$\overline{vy} = \overline{\overline{uxx}}^{-1} = u \quad .$$

Thus (b) holds in all cases but the last, which yields (a).

(iv) Let $\overline{uxux}^{-1} = \overline{vyvy}^{-1}$, $ux, vy \in UT\backslash U$; then

$$\overline{xux}^{-1}\overline{vyy}^{-1} = u^{-1}v \quad . \tag{4}$$

Now by parts (i) and (ii), $\overline{vyy}^{-1}$ does not belong to U , so we can apply part (iii) to the left hand side of (4), which is thus equal to either (a) e , whereupon $u = v$ and $x = y$, or (b) a reduced word of the form $x...y^{-1}$ so that $u^{-1}v \neq e$, and reducing the right hand side, either u ends in x^{-1} whence $ux \in U$, or v ends in y^{-1} whence $vy \in U$ (using Lemma 2) and both these possibilities are excluded. This proves that the $\overline{uxux}^{-1}$ are all distinct, provided $ux \notin U$.

Letting

$$\hat{B} = \{u\overline{xux}^{-1} \mid u \in U, x \in X^{-1}\}\backslash\{e\} \quad ,$$

the above implies that the set in question is $B \mathbin{\dot\cup} \hat{B}$. Since, for $u \in U$, $x \in T$,

$$(u\overline{xux}^{-1})^{-1} = \overline{uxx}^{-1}u^{-1} = \overline{uxx}^{-1}\overline{uxx}^{-1^{-1}} \quad ,$$

by part (ii), we have that $B^{-1} \subseteq \hat{B}$ and $\hat{B}^{-1} \subseteq B$, so that $\hat{B} = B^{-1}$, as required.

5. The main theorem.

Theorem 1 (Nielsen-Schreier). *If* F *is free and* H *is a subgroup of* F *, then* H *is free. If* $|F:H| = g$ *and the rank* r *of* F *are both finite, then the rank of* H *is equal to* $(r-1)g + 1$ *.*

16

Proof. We prove, using the above notation, that H is free on the set B . Since $A = B \mathbin{\dot\cup} B^{-1} \mathbin{\dot\cup} \{e\}$ (Lemma 4, part (iv)), and A generates H (Lemma 3), we see at once that B generates H . Now let $b_1 \ldots b_n$, $n \geq 1$, be a reduced word in the elements of $B \mathbin{\dot\cup} B^{-1} = A \backslash \{e\}$; suppose that

$$b_i = u_i x_i \overline{u_i x_i}^{-1} \, , \quad 1 \leq i \leq n \, ,$$

$u_i \in U$, $x_i \in T$, $u_i x_i \notin U$, and consider the product

$$b_i b_{i+1} = u_i x_i \overline{u_i x_i}^{-1} u_{i+1} x_{i+1} \overline{u_{i+1} x_{i+1}}^{-1} \, ,$$

for some i between 1 and $n-1$. Since $b_1 \ldots b_n$ is reduced, $b_i b_{i+1} \neq e$, and so by Lemma 4(iii), $x_i \overline{u_i x_i}^{-1} u_{i+1} x_{i+1}$ is equal to a reduced word in the elements of T of the form $x_i \ldots x_{i+1}$, of length at least two, and so

$$b_1 \ldots b_n = \ldots x_1 \ldots x_2 \ldots \; \ldots \; \ldots x_n \ldots \, ,$$

the right hand side being a reduced word in T which has length at least $n \geq 1$. Hence, $b_1 \ldots b_n \neq e$, and we have proved that there is no non-trivial relation in H between the elements of B . Theorem 1.3 implies that H is free on B .

6. Assume from now on that the rank r of F is finite. We first prove the numerical part of the theorem in the case when H is a normal subgroup N of F . Let N have finite index k in F . We modify the construction of the Schreier transversal by restricting our attention to reduced words $x_1 \ldots x_n$ in F with each x_i in X (rather than in T). If such words are called positive, we choose as the representative in U of any right coset to be the least *positive* word in that coset (for this U , note that e represents the coset N). It is easy to check that the Schreier property holds (see Exercise 3 below), but we must first show that each non-trivial coset does in fact contain a positive element, and this we now do. If $x \in T$, then

$Nx \in F/N = G$ say, a group of order k , so that by Lagrange's Theorem, $(Nx)^k = N$, that is, $x^k \in N$. Now let $x_1 \ldots x_n$ be any reduced word in T ; then for all i with $x_i^{-1} \in X$,

$$Nx_1 \ldots x_n = x_1 \ldots x_{i-1}(Nx_i)x_{i+1} \ldots x_n \text{ , since } N \text{ normal,}$$

$$= x_1 \ldots x_{i-1}(Nx_i^{-k}x_i)x_{i+1} \ldots x_n \text{ , as } x_i^k \in N \text{ ,}$$

$$= Nx_1 \ldots x_{i-1}x_i^{-k+1}x_{i+1} \ldots x_n \text{ ,}$$

and $x_i^{-k+1} = (x_i^{-1})^{k-1}$, a reduced word in X . Performing this operation for each i with $x_i^{-1} \in X$, we obtain

$$Nx_1 \ldots x_n = Nw \text{ ,}$$

where w is a reduced word in X (rather than T), so that $w \in Nx_1 \ldots x_n$, as required.

By the remark at the end of Step 2 (above), N is freely generated by the set B constructed from this new U . Now consider the elements

$$ux, u \in U, x \in X \text{ .}$$

Since the $\overline{ux}ux^{-1} \neq e$ are all distinct by Lemma 4(iv), and there are kr of them altogether, we must show that precisely $k-1$ of them are e , that is, precisely $k-1$ of the ux belong to U . If $v = x_1 \ldots x_n \in U\backslash\{e\}$ (so that all $x_i \in X$, $n \geq 1$), then $v = ux_n$ with $u \in U$. So every element of $U\backslash\{e\}$ appears in the set of ux's , that is, $UX \cap U = U\backslash\{e\}$, as required.

7. Now let $H \leq F$ be arbitrary of finite index g . Let C be the set of g right cosets of H in F , and for each $w \in F$ let

$$\left. \begin{array}{l} \tau_w : C \to C \\ \\ Hv \mapsto Hvw \end{array} \right\} \text{ .}$$

As each τ_w is one-to-one, we get a mapping

$$\left.\begin{array}{l} \tau \,:\, F \to S_g \\[1em] \quad w \mapsto \tau_w \end{array}\right\}$$

of F into the symmetric group of degree g, which is obviously a homomorphism. Ker τ is thus a normal subgroup of F contained in H and having index at most $g!$. Put $N = \text{Ker }\tau$, so that

$$N \lhd F, \; N \leq H \text{ (so } N \lhd H), \; |F : N| < \infty \quad .$$

Let $|H : N| = h$, so that $|F : N| = |F : H|\,|H : N| = gh$. Then, by the normal case proved above, with $r(H) = \text{rank of } H$, $r(N) = \text{rank}$ of N, we have:

$$N \lhd F, \; |F : N| = gh \Rightarrow r(N) = (r-1)gh + 1 \quad , \quad \text{and}$$

$$N \lhd H, \; |H : N| = h \Rightarrow r(N) = (r(H)-1)h + 1 \quad ,$$

whence $r(H) = (r-1)g + 1$, as required.

It would be both instructive and useful at this point to work through the various steps of the proof of the theorem in the case of a particular example, but as this can be done both more easily and more profitably using the idea of a group presentation, we postpone our example until the next section.

Remark. In conclusion, it is appropriate to say a few words about Nielsen's original proof of the subgroup theorem. Given a subset $Y = \{y_i \mid i \in I\}$ of a free group $F = F(X)$, it is transformed into another by any of the following three operations:

i'	replace y_i by y_i^{-1} ,	
ij	replace y_i by $y_i y_j$,	
\i	remove y_i , if $y_i = e$,	

where $i, j \in I$, $i \neq j$, and in each case all y_k are unaffected
for $k \neq i$. These operations are called *elementary Nielsen
transformations* and are analogous with the elementary row and
column operations on matrices used to prove the corresponding subgroup theorem in the abelian case (see Dedekind's theorem in §6
below). A finite sequence of such moves is called a *Nielsen
transformation*, and this is *regular* if no move of type \i is
involved. Given a Nielsen transformation τ of a subset $Y \subseteq F$,
we note the following facts:

   (i) $\langle Y \rangle = \langle Y\tau \rangle$,

   (ii) if τ is regular, there is a regular Nielsen transformation from $Y\tau$ to Y ,

   (iii) any permutation of any finite subset of Y can be
achieved by a regular Nielsen transformation,

   (iv) if τ is regular and Y is a basis for F , then so
is $Y\tau$.

The first of these is obvious and the others are exercises.

   A subset $Y \subseteq F$ is called N-*reduced* if, for any $x, y, z \in Y^{\pm 1}$,
the following three conditions hold:

   $x \neq e$,

   $xy \neq e \Rightarrow \ell(xy) \geq \ell(x), \ell(y)$,

   $xy \neq e \neq yz \Rightarrow \ell(xyz) > \ell(x) - \ell(y) + \ell(z)$,

where for $w \in F$, $\ell(w)$ denotes the length of w as a reduced
word in $X^{\pm 1}$. The crux of Nielsen's proof lies in establishing
the following assertion:

(*) $\forall$ finite $Y \subseteq F$, $\exists$ Nielsen transformation τ such that $Y\tau$
is N-reduced.

This requires a delicate induction, and we omit it. On the other
hand, it is a relatively simple matter to deduce from (*) that
finitely-generated subgroups of free groups are free. The proof
of (*) can be adapted to cover the case when Y is infinite, and

the whole approach has some valuable by-products (see Exercises 12-17).

EXERCISE 1. If F is a free group of rank ω , what is the cardinality of F as a set?

EXERCISE 2. If F is free of finite rank r , what is the number of words of length n in F ?

EXERCISE 3. Let F be free on a finite set X , and let N be a normal subgroup of F of finite index, so that each coset of N in F contains a positive word. Let U be the transversal consisting of the least *positive* word in each coset. Prove the following two-sided version of the Schreier property:

$$x_1 \ldots x_n \in U \Rightarrow x_1 \ldots x_{n-1} \ and \ x_2 \ldots x_n \in U \ ,$$

where each $x_i \in X$ and $n \geq 1$.

EXERCISE 4. Write each member of the symmetric group S_4 as a positive word in $x = (1234)$ and $y = (12)$ in such a way that the resulting set of 24 words satisfies the two-sided Schreier property of the previous exercise.

EXERCISE 5. Let H be a subgroup of an *arbitrary* group G with $|G:H| = g < \infty$. Prove that if G can be generated by r elements, then H can be generated by $(r-1)g + 1$ elements.

EXERCISE 6. Let H be a subgroup of a group G with $|G:H| = g < \infty$. Prove that G has a normal subgroup N such that $g \leq |G:N| \leq g !$.

EXERCISE 7. Let H be a subgroup of finite index in a free group $F = F(X)$ with Schreier transversal U . Let Γ be the graph whose vertices P_u are in one-to-one correspondence with U , and with an edge labelled $x \in X$ from P_u to P_v if and

only if $ux = v$. Prove that Γ is a tree (no loops), and use
Euler's formula to deduce that the set $\{(u,x) \in U{\times}X \mid ux \in U\}$ has
exactly $|U| - 1$ elements.

EXERCISE 8. Let F be free on X , and let X' be a set obtained from X by either
    a) replacing $x \in X$ by x^{-1} , or
    b) replacing $x \in X$ by xy $(x \neq y \in X)$,
and leaving all other elements of X fixed. Prove that in either
case, X' is a basis for F .

EXERCISE 9. If Y is a subset of a free group F and τ is a
regular Nielsen transformation of Y , prove that there is a regular Nielsen transformation from $Y\tau$ back to Y .

EXERCISE 10. Prove that any permutation of a finite subset of Y
can be effected by a regular Nielsen transformation.

EXERCISE 11. Let Y be a Nielsen reduced subset of a free group
$F = F(X)$, and let $w = y_1 {\dots} y_m$, $m \geq 0$, $y_i \in Y^{\pm 1}$ and $y_i y_{i+1} \neq e$
for all possible i . Prove that, as a word in $X^{\pm 1}$, $\ell(w) \geq m$.
Deduce that $<Y>$ is free on Y , and use (*) to prove that
finitely-generated subgroups of free groups are free.

EXERCISE 12. Let F be free on X and let $Y \subseteq F$ be N-reduced.
Prove that $X^{\pm 1} \cap <Y> = X^{\pm 1} \cap Y^{\pm 1}$. If in addition Y is a basis
for F , prove that $X^{\pm 1} = Y^{\pm 1}$. Use this in conjunction with (*)
to obtain information about $\text{Aut } F$; prove, for example, that if
F is finitely generated, then so is $\text{Aut } F$, and write down some
relations that hold between the generators you have constructed.

EXERCISE 13. If F is a free group of finite rank r , use (*)
to prove that F cannot be generated by fewer than r elements.
Show further that if a set Y of r elements generates F ,
then it is a basis for F . Deduce that free groups of finite
rank are *Hopfian*, that is, if $N \triangleleft F$ and $F/N \cong F$, then $N = E$.

EXERCISE 14. If x, y are distinct elements of a basis of a free group, prove that the set $\{y^{-n}xy^n \mid n \in Z\}$ is N-reduced. Deduce that a free group of finite positive rank is isomorphic to a proper subgroup of itself (in contrast to the conclusion of the previous exercise).

EXERCISE 15. Let ϕ be a homomorphism from a finitely-generated free group F onto a free group G. Prove that F has a basis $S = S_1 \overset{.}{\cup} S_2$ such that ϕ maps $<S_1>$ isomorphically onto G and $<S_2>$ to E.

EXERCISE 16. If a and b are elements of a free group such that a^m commutes with b^n $(m, n \in Z\backslash\{0\})$, prove that a and b are powers of a common element c.

EXERCISE 17. Use the previous exercise (thrice!) to show that the relation: $a \sim b \iff ab = ba$, defined on the non-trivial elements of a free group F, is an equivalence relation. Deduce that if $e \neq a \in F$, the centralizer $C_F(a) = \{w \in F \mid aw = wa\}$ is cyclic.

§3. Free presentations of groups

Suppose that

X is a set,

$F = F(X)$ is the free group on X,

R is a subset of F,

$N = \bar{R}$ is the normal closure of R in F, and

G is the factor group F/N.

Definition 1. With this notation we write $G = <X \mid R>$, and call this a *free presentation*, or simply a *presentation*, of G. The elements of X are called *generators*, and those of R *relators*. A group G is called *finitely presented* if it has such a presentation with both X and R finite sets.

Theorem 1.   *Every group has a presentation, and every finite*
group is finitely presented.

Proof.   If $G = \langle X \rangle$ and $\theta : F(X) \twoheadrightarrow G$ is the epimorphism of
Theorem 1.4, then $G \cong \langle X \mid \mathrm{Ker}\ \theta \rangle$. If $|G| = g < \infty$, we can take
$X = G$ and replace $\mathrm{Ker}\ \theta$ by a set of free generators for $\mathrm{Ker}\ \theta$.
This is a presentation with g generators and $g^2 - g + 1$ re-
lators, by the Nielsen-Schreier theorem.

Straight from the definition, we see that the free group on a
set X has the presentation $F = \langle X \mid\ \rangle$; in fact, it follows
from Theorem 1.3 that $\langle X \mid R \rangle$ is free on X if and only if $R = \emptyset$
or $\{e\}$. By convention, the group $\langle \emptyset \mid \emptyset \rangle$ is trivial (free of
rank 0). Taking the case $X = \{x\}$ and $R = \{x^n\}$, $n \in \mathbb{N}$, it
follows from Remark 1.4 and elementary group theory that the resulting group is just the multiplicative version of the additive
group of integers modulo n (since for a subset S of an abelian
group, $\bar{S} = \langle S \rangle$).

Theorem 2.   (i) *If* F *is free of rank* $n \geq 0$ *, we have*

$$F = \langle X \mid\ \rangle\ ,$$

where X *is a set with* $|X| = n$. (ii) *For the cyclic group of*
order $n \in \mathbb{N}$ *, we have*

$$Z_n = \langle x \mid x^n \rangle\ .$$

In view of the latter assertion, the sanguine reader might
reasonably suspect the group

$$G = \langle x, y \mid x^3, y^2, [x,y] \rangle\ ,$$

where $[x,y]$ denotes the commutator $x^{-1}y^{-1}xy$, to be none other
than the direct product $Z_2 \times Z_3$ (alias Z_6). Nor would he be

24

disappointed (Theorem 4.2 below). Thus we arrive at a suitably non-trivial example with which to illustrate the proof of the Nielsen-Schreier Theorem.

Example 1. Let us take

$$F = \langle x,y| \; \rangle, \; R = \{x^3, y^2, [x,y]\}, \; \bar{R} = N, \; G = F/N,$$

so we are armed with the prior knowledge that $|F : N| = 6$ (see Exercise 1).

Our first task is to find the Schreier transversal for N in F, modulo the ordering $x < y$ on $\{x,y\}$. The first seventeen elements of F are thus

$$e, \; x, \; y, \; x^{-1}, \; y^{-1}, \; x^2, \; xy, \; xy^{-1}, \; yx, \; y^2, \; yx^{-1}, \; x^{-1}y,$$
$$x^{-2}, \; x^{-1}y^{-1}, \; y^{-1}x, \; y^{-1}x^{-1}, \; y^{-2} \; .$$

Only six of these elements, namely

$$e, \; x, \; y, \; x^{-1}, \; xy, \; yx^{-1} \tag{1}$$

can belong to our Schreier transversal U, because

$$y^{-1} = y^{-2}y \in Ny, \quad \text{and} \quad y < y^{-1} \; .$$

This also excludes $xy^{-1}, \; x^{-1}y^{-1}, \; y^{-1}x, \; y^{-1}x^{-1}, \; y^{-2}$, since these all involve y^{-1} (and $N \trianglelefteq F$). Further,

$$x^2 = x^3x^{-1} \in Nx^{-1} , \quad yx = xy[x,y]^{-1} \in xyN = Nxy ,$$
$$y^2 = y^2e \in Ne , \quad x^{-1}y = y([x,y]^{-1})y^{-1}yx^{-1} \in Nyx^{-1} ,$$
$$x^{-2} = (x^3)^{-1}x \in Nx \; .$$

We now claim that U contains no element of length greater than 2, for if the reduced word $x_1 \ldots x_n \in U$, $n > 2$, each

$x_i \in \{x, y, x^{-1}, y^{-1}\}$, then $x_1 x_2 \in U$ and $x_1 x_2 x_3 \in U$, by the Schreier property. So, by (1),

$$x_1 x_2 = xy \quad \text{or} \quad yx^{-1} \quad,$$

and so $x_1 x_2 x_3$ must be one of:

$$xyx, \ xy^2, \ xyx^{-1}, \ yx^{-2}, \ yx^{-1}y, \ yx^{-1}y^{-1} \quad.$$

Now

$$xyx \in Nxyx = Nyx^2 = yNx^2 = yNx^{-1} = Nyx^{-1} \quad;$$

and similar manipulation yields that the remaining five words lie in

$$Nx, \ Ny, \ Nxy, \ Nx^{-1}, \ Nx^{-1}$$

respectively, contradicting the fact that $x_1 x_2 x_3 \in U$. U thus contains no element other than those in (1), and as $|U| = |G : N|$ $= 6$, these represent distinct cosets.

We obtain the sets A and B by means of the following table.

U \ T	x	y	x^{-1}	y^{-1}
e	e	e	e	y^{-2}
x	x^3	e	e	$xy^{-2}x^{-1}$
y	$yxy^{-1}x^{-1}$	y^2	e	e
x^{-1}	e	$x^{-1}yxy^{-1}$	x^{-3}	$x^{-1}y^{-1}xy^{-1}$
xy	xyx^2y^{-1}	xy^2x^{-1}	$xyx^{-1}y^{-1}$	e
yx^{-1}	e	$yx^{-1}yx$	$yx^{-2}y^{-1}x^{-1}$	$yx^{-1}y^{-1}x$

The element in the u-row and t-column is $\overline{ut}\overline{ut}^{-1}$. The entries in the table yield A . The non-e elements in the left-hand part of the table yield B , while those on the right give B^{-1} . We list these elements in two columns:

$$x^3 \qquad\qquad x^{-3} \qquad\qquad r$$

$$yxy^{-1}x^{-1} \qquad xyx^{-1}y^{-1} \qquad (t^{-1})^{y^{-1}x^{-1}}$$

$$y^2 \qquad\qquad y^{-2} \qquad\qquad s$$

$$x^{-1}yxy^{-1} \qquad yx^{-1}y^{-1}x \qquad (t^{-1})^{y^{-1}}$$

$$xyx^2y^{-1} \qquad yx^{-2}y^{-1}x^{-1} \qquad t^{x^{-1}y^{-1}}r^{y^{-1}}$$

$$xy^2x^{-1} \qquad xy^{-2}x^{-1} \qquad s^{x^{-1}}$$

$$yx^{-1}yx \qquad x^{-1}y^{-1}xy^{-1} \qquad st^{-1}$$

where the third column gives the elements of B as products of conjugates of $r = x^3$, $s = y^2$, $t = [x,y]$, and their inverses.

Note that $|B| = 7 = (2-1)6 + 1$, as required. We leave the details of finding a *positive* Schreier transversal and the corresponding set of free generators of N as an exercise, merely remarking that they are

$$U = \{e, x, y, x^2, xy, x^2y\} \quad,$$

$$B = \{yxy^{-1}x^{-1}, y^2, x^3, xyxy^{-1}x^{-2}, xy^2x^{-1}, x^2yxy^{-1}, x^2y^2x^{-2}\} \quad,$$

respectively.

We now go on to prove a simple but tremendously useful result (Theorem 4) about presentations, that will form the basis for some key ideas in the next two sections.

Lemma 1. *Let* X, Y, Z *be groups, and* $\alpha : X \to Y$, $\beta : X \to Z$ *homomorphisms with* α *onto and such that* $\operatorname{Ker} \alpha \subseteq \operatorname{Ker} \beta$. *Then there is a homomorphism* $\gamma : Y \to Z$ *such that* $\alpha\gamma = \beta$.

Proof. We have the following diagram, where all the unlabelled maps are inclusions:

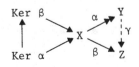

Now for any $y \in Y$, *choose* $x \in X$ such that $x\alpha = y$, and define

$$y\gamma = x\beta \ . \tag{2}$$

Our first task is to show that this is well defined. Thus, let $x' \in X$ be another pre-image of y, so that $x'\alpha = y = x\alpha$. Then

$$x'x^{-1} \in \text{Ker } \alpha \subseteq \text{Ker } \beta \ ,$$

so that

$$(x'x^{-1})\beta = e \ , \quad \text{whence} \quad x'\beta = x\beta \ .$$

Thus the definition of $y\gamma$ is independent of the choice of x. For any $x \in X$, x is a pre-image under α of $x\alpha$, so that by (2),

$$x(\alpha\gamma) = (x\alpha)\gamma = x\beta \ .$$

It remains only to show that γ is a homomorphism. Let $y_1, y_2 \in Y$, and let $x_1, x_2 \in X$ be pre-images of them. Then

$$(x_1 x_2)\alpha = x_1 \alpha x_2 \alpha \ , \quad \text{since} \quad \alpha \text{ is a homomorphism,}$$
$$= y_1 y_2 \ ,$$

so that $x_1 x_2$ is a pre-image of $y_1 y_2$. Thus by (2),

$$(y_1 y_2)\gamma = (x_1 x_2)\beta$$
$$= x_1 \beta x_2 \beta \ , \quad \text{since} \quad \beta \text{ is a homomorphism,}$$
$$= y_1 \gamma y_2 \gamma \ ,$$

completing the proof.

Theorem 3 (von Dyck). *If* R *and* S *are subsets of the free*

group F on a set X such that $R \subseteq S$, then there is an epimorphism

$$\theta : <X|R> \to <X|S>$$

which fixes X elementwise. The kernel of θ is just the normal
closure of $S\backslash R$ as a subset of $<X|R>$.

Proof. The first assertion is a simple application of the lemma,
with α and β the natural maps

$$F \twoheadrightarrow F/\bar{R} , \quad F \twoheadrightarrow F/\bar{S} ,$$

respectively. Since α is onto and $\alpha\theta = \beta$, we have:

$$\text{Ker } \theta = (\text{Ker } \beta)\alpha = \bar{S}\alpha = \overline{S\alpha} = \overline{R\alpha \cup (S\backslash R)\alpha} ,$$

and since $R \subseteq \text{Ker } \alpha$, $\text{Ker } \theta = \overline{(S\backslash R)\alpha}$, as claimed.

The perceptive reader will have spotted our appeal, in the statement of this theorem, to a technique known as san (systematic
abuse of notation). This phenomenon is endemic and ineradicable,
and arises in the following way. Underlying any presentation
$G = <X|R>$, there is a free group $F = <X| >$ consisting of reduced words w in $X \cup X^{-1}$. We like to think of G as being
generated by X and containing elements such as w , whereas in
reality G consists of *cosets* of the form $\bar{R}w$. We hope that
what is meant will always be clear from the context, and that any-
one who remains unconvinced of the need for this technique will
write out sanfree proofs of (say) Theorem 3 (above) and Theorem 4.2
(below).

Theorem 4 (Substitution Test). *Suppose we are given a presen-
tation* $G = <X|R>$ *, a group* H *and a mapping* $\theta : X \to H$ *. Then*
θ *extends to a homomorphism* $\theta'' : G \to H$ *if and only if, for all*

$x \in X$ *and all* $r \in R$ *, the result of substituting* $x\theta$ *for* x *in* r *yields the identity of* H *.*

Proof. Let $F = <X| >$ and consider the commutative diagram:

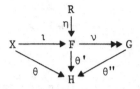

where η and ι are inclusions and ν is the natural homomor-phism. Since F is free on X , θ extends (uniquely) to a homomorphism $\theta' : F \to H$, and our substitution condition can be rephrased simply as: $R \subseteq \text{Ker } \theta'$. Since $\text{Ker } \theta' \lhd F$ and $\bar{R} = \text{Ker } \nu$, this condition is equivalent to $\text{Ker } \nu \le \text{Ker } \theta'$. The existence of $\theta'' : G \to H$ extending θ is a consequence of Lemma 1. For the converse, the existence of such a θ'' entails that

$$R \subseteq \bar{R} = \text{Ker } \nu \subseteq \text{Ker } \nu\theta'' = \text{Ker } \theta' \quad .$$

Note that when such a θ'' exists it must be unique since X generates G . Further, if H is generated by the $x\theta$, θ'' must be onto and $|H| \le |G|$. This will be useful later on in the quest for presentations of specific concrete groups.

EXERCISE 1. Let $G = <X|R>$, where

$$X = \{x,y\} \quad, \quad R = \{x^3, y^2, [x,y]\} \quad,$$

and let $\theta : X \to Z_6 = <a|a^6>$ be given by $x\theta = a^2$, $y\theta = a^3$. If θ' is the corresponding extension to $F = <x,y| >$, prove that $\text{Ker } \theta' \supseteq R$ and $\text{Im } \theta' = Z_6$. Deduce that $|G| \ge 6$.

EXERCISE 2. Let $F = <x,y| >$ and $N = \{x^3, y^2, [x,y]\} \lhd F$, as in Example 1 and Exercise 1. Write down a *positive* Schreier trans-

versal for N in F and use it to obtain a set of free generators for N .

EXERCISE 3. Let $G = \langle X|R \rangle$, where

$$X = \{x,y\} \ , \quad R = \{x^3, y^2, (xy)^2\} \ ,$$

and let $\theta : X \to S_3$ be given by $x\theta = (123)$, $y\theta = (12)$. As in
Exercise 1, deduce that $|G| \geq 6$, and go through the proof of
the Neilsen-Schreier theorem with this $\bar{R}$ in the role of H .

EXERCISE 4. Prove that if G has a presentation $\langle x,y \mid R \rangle$,
then G/G' is presented by $\langle x,y \mid R, [x,y] \rangle$. Can you extend this
result to the case of more than two generators?

EXERCISE 5. Consider that group $F(2,6)$ given by

$$\langle x_1, x_2, x_3, x_4, x_5, x_6 \mid x_1 x_2 x_3^{-1}, x_2 x_3 x_4^{-1}, x_3 x_4 x_5^{-1}, x_4 x_5 x_6^{-1},$$
$$x_5 x_6 x_1^{-1}, x_6 x_1 x_2^{-1} \rangle \ .$$

Prove that there is a homomorphism $\chi : F(2,6) \to S_\infty$ such that

$$n(x_1 \chi) = n + 1 \ , \quad n(x_2 \chi) = -n \ , \quad n \in Z \ ,$$

and deduce that this group is infinite.

EXERCISE 6. Prove that, for any non-negative integer r , the
subgroup

$$Q_r = \{a/r^i \mid i, a \in Z \ , \ i \geq 0\}$$

of the additive group Q of rational numbers has the presentation

$$Q_r = \langle x_0, x_1, \ldots, x_n, \ldots \mid rx_i / x_{i-1} \ , \ \forall \ i \in N \rangle \ .$$

EXERCISE 7. Regarding the additive group Z of integers as a subgroup of Q and each Q_r (see Exercise 6), prove that Q/Z is isomorphic to the direct sum of the groups Q_p/Z , where p runs over the primes.

§4. Elementary properties of presentations

We collect together in this section three important applications of the Substitution Test, the first of which gives an alternative proof of Theorem 3.1.

Theorem 1. *If* $m : G \times G \to G$ *denotes the binary operation on a group* G *, then* G *has the presentation* <X|R> *, where* X *is the underlying set of* G *and* $R = \{xym(x,y)^{-1} \mid x,y \in G\}$.

Proof. Letting M be the group <X|R> , the identity map 1_G extends to a homomorphism $\alpha : M \to G$ by the Substitution Test. Letting $\beta : G \to M$ be the 'inclusion' mapping, it is clear that both $\alpha\beta$ and $\beta\alpha$ fix G . They are thus equal to 1_M and 1_G respectively, proving that α is an isomorphism.

Theorem 2. *If* $G = <X|R>$ *and* $H = <Y|S>$ *are two presentations, then the direct product* $G \times H$ *has the presentation*

$$<X,Y \mid R,S , [X,Y]> , \tag{1}$$

where [X,Y] *denotes the set of commutators* $\{[x,y] \mid x \in X, y \in Y\}$.

Proof. Let D denote the group presented by (1). By the Substitution Test, the 'inclusions'

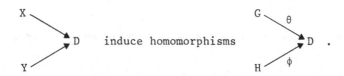

induce homomorphisms

(The inverted commas are used in deference to the tradition that inclusions be one-to-one.) The relators [X,Y] guarantee that the images of θ and ϕ centralize one another in D , whence we obtain a homomorphism

$$\left.\begin{aligned} \alpha \;:\; & G \times H \;\to\; D \\ & (g,h) \;\mapsto\; g\theta h\phi \end{aligned}\right\} \;,$$

which fixes $X \cup Y$ elementwise. On the other hand, the 'inclusion' of $X \cup Y$ in $G \times H$ extends (by the Substitution Test again) to a homomorphism $\beta : D \to G \times H$ with the same property. As in the previous proof, we conclude that α is an isomorphism.

With $G = <X|R>$ and $H = <Y|S>$, the inquisitive reader might be tempted to raise the question of the significance of the group $P = <X,Y \mid R,S>$, obtained from G and H by an apparently even more natural rule than $G \times H$. Before answering this, we pose another question, namely, given groups G and H , does there exist a group D and homomorphisms $\alpha_1 : D \to G$, $\alpha_2 : D \to H$ such that, for any group K and any homomorphisms $\beta_1 : K \to G$, $\beta_2 : K \to H$, there is a unique homomorphism $\gamma : K \to D$ such that $\gamma\alpha_1 = \beta_1$, $\gamma\alpha_2 = \beta_2$? This is an example of a universal property (cf. the definition of free groups), and is best reduced to diagrammatic form:

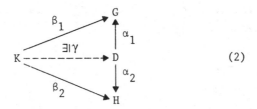

$$(2)$$

We leave it as an exercise to show that the direct product $D = G \times H$ is the unique answer to this question. On the other hand, the group P is the unique answer to the *dual* question, as illustrated by the following diagram:

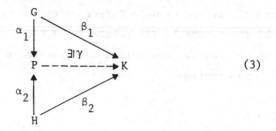

$$(3)$$

We confine both the formulation of, and the answer to, this question to the exercises.

Definition.   If $G = \langle X|R\rangle$ and $H = \langle Y|S\rangle$, then the group $\langle X,Y \mid R,S\rangle$ is called the *free product* of G and H , and is denoted by $G * H$. The kernel of the natural homomorphism from $G * H$ to $G \times H$ is called the *Cartesian*.

The general theory of free products forms an important part of combinatorial group theory. Though it is (sadly) beyond the scope of these notes, we shall run across one or two interesting examples of free products later on (see §§25 and 30, for example).

We pass on to our final application of Theorem 3.4 which, though it may appear trifling, none the less forms the basis of a very valuable practical tool.

Theorem 3.   *Let* $F = \langle X| \;\rangle$, $G = \langle X|R\rangle$ *and suppose that* $w,r \in F$ *with* w *arbitrary and* $r \in \bar{R}\backslash R$. *If* y *is a symbol not in* X , *then both the 'inclusions'*

$$X \rightarrow \langle X \mid R,r\rangle$$
$$X \rightarrow \langle X,y \mid R,y^{-1}w\rangle$$

extend to isomorphisms with domain G .

Proof.   As in the proofs of the previous two theorems, this is a straightforward application of the Substitution Test. Of the four mappings involved, three merely fix X elementwise, while

34

the fourth in addition sends y to w , regarded as a member of
G by san.

The four isomorphisms of Theorem 3 yield four ways of adjusting
a given presentation $<X|R>$ to obtain another, $<X'|R'>$ say, of
the same group. These are called the *Tietze transformations* and
are defined as follows, where $F = <X| >$ throughout.

R+ , adjoining a relator:

$$X' = X , \quad R' = R \cup \{r\} ,$$

where $r \in \bar{R}\backslash R$ (normal closure in F).

R- , removing a relator:

$$X' = X , \quad R' = R\backslash\{r\} ,$$

where $r \in R \cap \overline{R\backslash\{r\}}$.

X+ , adjoining a generator:

$$X' = X \cup \{y\} , \quad R' = R \cup \{y^{-1}w\} ,$$

where $y \notin X$ and $w \in F$.

X- , removing a generator:

$$X' = X\backslash\{y\} , \quad R' = R\backslash\{y^{-1}w\} ,$$

where $y \in X$, $w \in <X\backslash\{y\}| >$ and $y^{-1}w$ is the only member of R
involving y .

Example 1.   Consider the von Dyck group

$$D(\ell,m,n) = <x,y \mid x^{\ell},y^{m},(xy)^{n}> ,$$

35

where ℓ, m, n are integers (usually positive). We apply a succession of Tietze transformations to this presentation in accordance with the scheme set out below. The generators are omitted since at each stage they are just the letters involved in the relators.

$$x^\ell\ y^m\ (xy)^n$$

X+ $\quad x^\ell\ y^m\ (xy)^n\ a^{-1}xy$

R+ $\quad x^\ell\ y^m\ (xy)^n\ a^{-1}xy\ a^n$

R- $\quad x^\ell\ y^m\ \qquad a^{-1}xy\ a^n$

R+ $\quad x^\ell\ y^m\ \qquad a^{-1}xy\ a^n\ (ay^{-1})^\ell$

R- $\quad\ \ y^m\ \qquad a^{-1}xy\ a^n\ (ay^{-1})^\ell$

R+ $\quad\ \ y^m\ \qquad a^{-1}xy\ a^n\ (ay^{-1})^\ell\ x^{-1}ay^{-1}$

R- $\quad\ \ y^m\ \qquad\qquad a^n\ (ay^{-1})^\ell\ x^{-1}ay^{-1}$

X- $\quad\ \ y^m\ \qquad\qquad a^n\ (ay^{-1})^\ell$

X+ $\quad\ \ y^m\ \qquad\qquad a^n\ (ay^{-1})^\ell\qquad b^{-1}y^{-1}$

R+ $\quad\ \ y^m\ \qquad\qquad a^n\ (ay^{-1})^\ell\qquad b^{-1}y^{-1}\ b^m$

R- $\qquad\qquad\qquad a^n\ (ay^{-1})^\ell\qquad b^{-1}y^{-1}\ b^m$

R+ $\qquad\qquad\qquad a^n\ (ay^{-1})^\ell\qquad b^{-1}y^{-1}\ b^m\ (ab)^\ell$

R- $\qquad\qquad\qquad a^n\ \qquad\qquad b^{-1}y^{-1}\ b^m\ (ab)^\ell$

R+ $\qquad\qquad\qquad a^n\ \qquad\qquad b^{-1}y^{-1}\ b^m\ (ab)^\ell\ y^{-1}b^{-1}$

R- $\qquad\qquad\qquad a^n\ \qquad\qquad\qquad\quad b^m\ (ab)^\ell\ y^{-1}b^{-1}$

X- $\qquad\qquad\qquad a^n\ \qquad\qquad\qquad\quad b^m\ (ab)^\ell$

This proves that $D(\ell,m,n)$ and $D(n,m,\ell)$ are isomorphic, though the method is rather long and tedious. A substantial simplification is achieved if we work in terms of *relations* rather than relators, where a (defining) relation is obtained from the corresponding relator by setting it equal to e. Conversely, if w_1 and w_2 are words in the generators, the relation $w_1 = w_2$ yields the relator $w_1^{-1}w_2$, for example. Thus, starting from

$$D(\ell,m,n) = \langle x,y \mid x^\ell = y^m = (xy)^n = e \rangle\ ,$$

the above computation may be informally paraphrased as follows.

Introduce the generator $a = xy$, so that

$$a^n = e , \quad x = ay^{-1} , \quad (ay^{-1})^\ell = e ,$$

and the presentation reduces to

$$\langle a,y \mid a^n = (ay^{-1})^\ell = y^m = e \rangle .$$

Replacing y^{-1} by b and inverting the last relation we obtain

$$\langle a,b \mid a^n = (ab)^\ell = b^m = e \rangle .$$

With a reasonable amount of care, no damage is done by this loss of precision. We give one more example, which will turn out to be useful later on.

Example 2.   Consider the group

$$TK = \langle x,y,z \mid x = yzy^{-1}, \ y = zxz^{-1}, \ z = xyx^{-1} \rangle .$$

Using the last relation to eliminate z , we see that the first two relations become equivalent and we have the group

$$B_3 = \langle x,y \mid xyx = yxy \rangle .$$

Putting $xy = a$, we have

$$y = x^{-1}a , \quad ax = x^{-1}a^2 .$$

Finally, letting $x = a^{-1}b$, we obtain the group

$$RM = \langle a,b \mid a^3 = b^2 \rangle .$$

Theorem 3 guarantees that the groups TK , B_3 , RM are all iso-morphic.

Our final result in this section complements Theorem 3, and is
surprisingly easy to prove.

Theorem 4. *Given two finite presentations of the same group,
one can be obtained from the other by a finite sequence of Tietze
transformations.*

Proof. Given two such presentations

$$<X|R(X) = e> \ , \quad <Y|S(Y) = e> \ ,$$

suppose that

$$X = X(Y) \ , \quad Y = Y(X) \tag{4}$$

are two systems of equations expressing the generators X in
terms of the generators Y , and vice versa. We now apply Tietze
transformations en bloc to the first presentation in accordance
with the following scheme.

```
X+ :  X,Y  R(X) = e, Y = Y(X),
R+ :  X,Y  R(X) = e, Y = Y(X), X = X(Y)
R+ :  X,Y  R(X) = e, Y = Y(X), X = X(Y), R(X(Y)) = e,
R- :  X,Y           Y = Y(X), X = X(Y), R(X(Y)) = e,
R+ :  X,Y           Y = Y(X), X = X(Y), R(X(Y)) = e, Y = Y(X(Y))
R- :  X,Y                     X = X(Y), R(X(Y)) = e, Y = Y(X(Y))
X- :    Y                               R(X(Y)) = e, Y = Y(X(Y))
R+ :    Y                               R(X(Y)) = e, Y = Y(X(Y)), S(Y) = e
R- :    Y                                                         S(Y) = e
```

Remark 1. In the course of this proof, we have added $|Y|$ gen-
erators and removed $|X|$, and R+, R- have been used

$$|X| + |R| + |Y| + |S| \ , \quad 2(|R| + |Y|)$$

38

times, respectively.

Remark 2.   In contrast to this theorem, there is no general algorithm for deciding whether two given finite presentations yield
isomorphic groups. (The apparent paradox is explained by the
fact that, in proving the theorem, we made implicit use of an
isomorphism to write down the equations (4).) This is known as
the Isomorphism Problem, and is the hardest of three problems
identified by M. Dehn. The other two are called the Word Problem
and the Conjugacy Problem and are also undecidable in general.
We shall have more to say sbout these problems later in §24.

Remark 3.   We warn the reader against a popular error. Given a
presentation $G = <X|R(X) = e>$ and new generators Y such that
$X = X(Y)$, it is in general false that G is presented in terms
of Y by $<Y|R(X(Y)) = e>$. The correct presentation on the new
generators may be culled from the proof of the theorem, and is

$$G = <Y|R(X(Y)) = e, Y = Y(X(Y))> .$$

EXERCISE 1.   In accordance with Theorem 1, use the multiplication
table of Z_3 to obtain a 3-generator, 9-relation presentation of
this group. Use Tietze transformations to show that the result is
in fact cyclic of order 3 .

EXERCISE 2.   Do you recognise the group

$$<a,b,c \mid a^3 = b^2 = c^2 = (ab)^2 = (bc)^2 = [a,c] = e> ?$$

What is the order of the element abc ?

EXERCISE 3.   Apply Theorem 2 inductively to find a presentation
of the direct sum of r copies of the infinite cyclic group Z
$(r \in N)$, and use Exercise 3.4 to deduce that this group is none
other than F/F' , where F is free of rank r .

EXERCISE 4.   If   G   and   H   are given groups, prove that the
direct product $G \times H$ and the free product $G * H$ are the
(unique) groups with the universal properties encapsulated in
diagrams (2) and (3), respectively. ,

EXERCISE 5.   Prove that the Cartesian of the free product
$D_\infty = Z_2 * Z_2$ is a cyclic group.

EXERCISE 6.   For any non-zero integers ℓ, m, n,   prove that

$$D(-\ell, m, n) \cong D(\ell, m, n) \cong D(m, \ell, n) \quad .$$

Deduce that $D(\ell, m, n)$ depends only on the absolute values of
ℓ, m, n and not on their order or signs.   Do you recognise the
group in the case $\ell = 1$?

EXERCISE 7.   Prove that the groups

$$\langle a, b, c, d \mid ab = c, \ bc = d, \ cd = a, \ da = b \rangle \quad ,$$
$$\langle a, b, c, d, Q, f \mid ab = d, \ bc = Q, \ cd = f, \ dQ = a, \ Qf = b, \ fa = c \rangle$$

are cyclic and find their orders.

EXERCISE 8.   Prove that the group

$$T = \langle a, b \mid a^b b^a = (b^{-1} a^2)^2 = e \rangle$$

is isomorphic to $Z_2 * Z_3$.

EXERCISE 9.   Find a presentation for $Z_6 = \langle x \mid x^6 \rangle$ in terms of
the generators $a = x^2$, $b = x^3$.

EXERCISE 10.   Check the commutator identity (known as the *Witt
identity*)

$$[[x, y], z^x][[z, x], y^z][[y, z], x^y] = e \quad ,$$

and use it to prove that the group

$$\langle x,y,z \mid [x,y] = z, [y,z] = x, [z,x] = y \rangle$$

is trivial.

EXERCISE 11.   Let x and y be members of a group G such that $x^{-1}y^n x = y^m$, $n,m \in Z$. Prove that, for any $k \in N$, $x^{-1}y^{n^k} x = y^{m^k}$.
Deduce that the group

$$\langle x,y \mid y^{-1}x^n y = x^{n+1}, \ x^{-1}y^n x = y^{n+1} \rangle$$

is trivial for any $n \in Z$.

2·Examples of presentations

> ... there is an Infinite in him, which with
> all his cunning, he cannot quite bury under
> the finite (Carlyle: Sartor Resartus).

We shall apply the results of the previous chapter to obtain
presentations for various familiar classes of concrete groups.
There is a general method which works in principle for all finite
groups and can be crystallized into four steps. Given a concrete
group G , proceed as follows.

1. Find a suitable set X of generators for G .
2. In terms of X , write down some relations R = e that hold
   in G and hope they are enough to define G .
3. Use concrete information to bound the order of G below, by
   b say.
4. Working with words in X , bound the order of <X|R> above,
   also by b .

By 1, 2 and the Substitution Test, we have an epimorphism

$$\theta \ : \ <X|R> \ \twoheadrightarrow \ G \ ,$$

and by 3, 4

$$b \leq |G| = |\text{Im } \theta| = |<X|R> : \text{Ker } \theta| \leq |<X|R>| \leq b \ \ ,$$

so that θ must be an isomorphism.

For finitely-generated abelian groups, we give enough of the
proof of the Basis theorem to write down presentations, and to
obtain a simple algorithm for describing G/G' in terms of any
finite presentation <X|R> of G . Another useful consequence

is the intuitively reasonable result that for $G = \langle X|R \rangle$ to be finite, we must have $|X| \leq |R|$.

§5. Some popular groups

When classifying groups of order 8 in a first course in group theory, the abelian groups $Z_2 \times Z_2 \times Z_2, Z_8, Z_2 \times Z_4$ and the dihedral group D_4 (see below) come out nicely in the wash and we are left with the following situation: the possibility of a group (or groups) G of order 8 containing an element x of order 4 and an element $y \notin \langle x \rangle$ such that $y^2 = x^2$ and $y^{-1}xy = x^{-1}$. There is at most one such group, since the given information implies that its elements must be

$$x^j y^k , \quad 0 \leq j \leq 3 , \quad 0 \leq k \leq 1 ,$$

with multiplication table

	x^ℓ	$x^\ell y$
x^j	$x^{j+\ell}$	$x^{j+\ell} y$
$x^j y$	$x^{j-\ell} y$	$x^{j-\ell+2}$

with powers of x, y reduced modulo 4, 2 respectively. To show that this does indeed define a group, one can either check the associative law directly, or appeal to a concrete group. We adopt the latter course and work out the example in detail to serve as a prototype.

Example 1. Let $GL(2,C)$ denote the group of all non-singular 2×2 complex matrices, and consider the subgroup Q generated by the matrices

$$\xi = \begin{pmatrix} i & 0 \\ 0 & -i \end{pmatrix} , \quad \eta = \begin{pmatrix} 0 & -1 \\ 1 & 0 \end{pmatrix} ,$$

where $i^2 = -1$. Now we have $|\xi| = 4$ and $\eta \notin \langle \xi \rangle$, so that

$|Q| \geq 8$. Furthermore, we have $\eta^2 = \xi^2$ and $\eta^{-1}\xi\eta = \xi^{-1}$, so we hopefully put

$$G = \langle x,y \mid x^4 = e, \; y^2 = x^2, \; y^{-1}xy = x^{-1} \rangle \tag{1}$$

and use the Substitution Test to obtain the epimorphism

$$
\left.
\begin{array}{l}
\theta \; : \; G \twoheadrightarrow Q \\[4pt]
\quad x \mapsto \xi \\[4pt]
\quad y \mapsto \eta
\end{array}
\right\}
$$

To show that θ is an isomorphism, we merely have to prove that $|G| \leq 8$.

To do this, it will be sufficient to show that any element of G is equal to a member of the set $U = \{x^i y^j \mid 0 \leq i \leq 3, \; 0 \leq j \leq 1\}$. Now it follows from the relations in (1) that U is closed under post-multiplication by $x^{\pm 1}$ and $y^{\pm 1}$. For example,

$$(x^i y)x = x^i y^2 y^{-1} xyy^{-1} = x^i x^2 x^{-1} y^{-1} = x^{i-1}y \; .$$

Thus, $Uw \subseteq U$ for any word w in $\{x^{\pm 1}, y^{\pm 1}\}$. Since these words cover G (by san) and $e \in U$, we have

$$G = eG \subseteq UG \subseteq U \; ,$$

as required.

The group given by (1) satisfies all the requirements for our proposed group of order 8 (above) and we have just shown that it is isomorphic to the concrete Q - called the *quaternions*. This completes the classification of groups of order 8 .

Definition 1. For n an integer ≥ 2 , we define the *generalized quaternion group* Q_{2n} to be the subgroup of $GL(2,C)$ generated by the matrices.

$$\xi = \begin{pmatrix} \omega & 0 \\ 0 & \bar{\omega} \end{pmatrix}, \quad \eta = \begin{pmatrix} 0 & -1 \\ 1 & 0 \end{pmatrix},$$

where $\omega = e^{i\pi/n} \in \mathbb{C}$.

Theorem 1. *The generalized quaternion group* Q_{2n} *is of order* $4n$ *and has the presentation*

$$<x,y \mid x^n = y^2, \; y^{-1}xy = x^{-1}> . \tag{2}$$

Proof. Letting G be the group presented by (2), we see that $x^n = y^2$ commutes with y , and so

$$x^n = (x^n)^y = (x^y)^n = x^{-n} ,$$

that is, $x^{2n} = e$ in G . Thus, the quaternion group Q is just Q_4 . We proceed as in the above example, and observe that mapping x,y to ξ,η respectively yields a homomorphism from G to Q_{2n} . That $|Q_{2n}| \geq 4n$ follows from the fact that ω is a *primitive* $2n^{th}$ root of unity and $\eta \notin <\xi>$. Finally, the proof that $|G| \leq 4n$ differs from the above only in the choice of $U = \{x^i y^j \mid 0 \leq i \leq 2n-1, \; 0 \leq j \leq 1\}$.

Definition 2. For an integer $n \geq 3$, the group D_n of symmetries of a regular plane n-gon in $\mathbb{R}^3$ is called the dihedral group of degree n .

Theorem 2. *The dihedral group* D_n *is of order* $2n$ *and has the presentation*

$$<x,y \mid x^n = y^2 = (xy)^2 = e> .$$

Proof. We number the vertices of the n-gon consecutively $1,2,\ldots,n$ clockwise. Since rigid transformations preserve adjacency of vertices, any symmetry is determined by its effect on the vertices 1 and 2, and there are n possible positions for 1 and given this, two for 2. Hence, D_n has order at most $2n$. On the other hand, if we let ξ denote counter-clockwise rotation through $2\pi/n$ about the centre O of the n-gon in its own plane, and η rotation through π about O1, we see that the symmetries $\{\xi^i \eta^j \mid 0 \le i \le n-1,\ 0 \le j \le 1\}$ are all distinct. To prove this, we look at the corresponding permutations $(1\ 2\ldots n),(2\ n)(3\ n-1)\ldots$ of the vertices; the first has order n and the second is not a power of the first. Now we just proceed as in the above example, observing that, from the last two relations in the proposed presentation,

$$y^{-1}xy = yxy = x^{-1} \ ,$$

as before.

Theorem 3. *The symmetric group* S_n $(n \in \mathbb{N})$ *has the presentation*

$$G_n = \langle x_1,\ldots,x_{n-1} \mid R,S,T \rangle \ ,$$

where

$$R = \{x_i^2 \mid 1 \le i \le n-1\} \ ,$$
$$S = \{(x_i x_{i+1})^3 \mid 1 \le i < n-1\} \ ,$$
$$T = \{[x_i,x_j] \mid 1 \le i < j-1 < n-1\} \ .$$

Proof. Letting G_n denote the group so presented, we follow the plan suggested at the start of this chapter and proceed in four steps.

46

Step 1: To find suitable generators, note just that S_n is generated by the cycles, and that

$$(a_1 \dots a_\ell) = (a_1 a_2)(a_1 a_3) \dots (a_1 a_\ell) \quad .$$

Letting $x_i' = (i \ i{+}1)$ for $1 \le i \le n{-}1$, we have for $1 \le i < j \le n$:

$$(ij) = x_i' x_{i+1}' \dots x_{j-2}' x_{j-1}' x_{j-2}' \dots x_{i+1}' x_i' \quad ,$$

so that S_n is generated by the set $\{x_i' \mid 1 \le i \le n{-}1\}$.

Step 2: The rule

$$\left. \begin{array}{l} G_n \twoheadrightarrow S_n \\[4pt] x_i \mapsto x_i' \end{array} \right\}$$

yields a homomorphism (onto by Step 1) by the Substitution Test, since

R : x_i' is a transposition,

S : $x_i' x_{i+1}'$ is a 3-cycle,

T : x_i', x_j' are disjoint for $|i-j| \ge 2$.

Step 3: This is just the well-known fact that $|S_n| \ge n!$.

Step 4: We prove that $|G_n| \le n!$ by induction on n . When $n = 1$, G_n is the trivial group $< \mid >$ of order $1 \le 1!$, and when $n = 2$, $G_n = <x_1 \mid x_1^2>$ which is cyclic of order $2 \le 2!$. We thus assume that $n \ge 3$ and that $G_{n-1} \le (n{-}1)!$. Let H be the subgroup of G_n generated by $x_1, \dots, x_{n-2}$ and define

$$y_0 = e, \ y_i = x_{n-1} \dots x_{n-i}, \ 1 \le i \le n{-}1 \quad .$$

Consider the subset

$$A = \{hy_i \mid h \in H, 0 \le i \le n-1\} \ ,$$

of G_n ; we shall prove that $A = G_n$. The hard bit (and the crux of the whole proof) is to show that

$$Ax_j \subseteq A \ , \ 1 \le j \le n-1 \ ,$$

so we consider the product $hy_i x_j$ in six possible cases.

(i) $\quad i = 0, \ j < n-1 : hy_i x_j = hx_j \in Hy_0 \subseteq A$, as $j < n-1$.

(ii) $\quad i = 0, \ j = n-1 : hy_i x_j = hex_{n-1} = hy_1 \in Hy_1 \subseteq A$.

(iii) $\quad i > 0, \ j > n-i : hy_i x_j = h(x_{n-1} \cdots x_j x_{j-1} \cdots x_{n-i}) x_j$
$$\qquad\qquad = hx_{n-1} \cdots x_j x_{j-1} x_j \cdots x_{n-i}, \text{ by } T \ ,$$
$$\qquad\qquad = hx_{n-1} \cdots x_{j-1} x_j x_{j-1} \cdots x_{n-i}, \text{ by } R$$
$$\qquad\qquad\qquad\qquad\qquad\qquad \text{and } S \ ,$$
$$\qquad\qquad = (hx_{j-1}) x_{n-1} \cdots x_{n-i}, \text{ by } T \ ,$$
$$\qquad\qquad \in Hy_i \subseteq A \ .$$

(iv) $\quad i > 0, \ j = n-i : hy_i x_j = hx_{n-1} \cdots x_{n-i} x_{n-i}$
$$\qquad\qquad = hx_{n-1} \cdots x_{n-i+1}, \text{ by } R \ ,$$
$$\qquad\qquad = hy_{i-1} \in Hy_{i-1} \subseteq A \ .$$

(v) $\quad i > 0, \ j = n-i-1 : hy_i x_j = hx_{n-1} \cdots x_{n-i} x_{n-(i+1)}$
$$\qquad\qquad = hy_{i+1} \in Hy_{i+1} \subseteq A \ .$$

(vi) $\quad i > 0, \ j < n-i-1 : hy_i x_j = (hx_j) y_i, \text{ by } T \ ,$
$$\qquad\qquad \in Hy_i \subseteq A \ .$$

This shows that $hy_i x_j \in A$, for all $h \in H$, $0 \le i \le n-1$ and $1 \le j \le n-1$, that is, that $Ax_j \subseteq A$, for all j . Because of R , we have

$$Ax_j^{-1} = Ax_j \subseteq A, \ 1 \le j \le n-1 \ ,$$

and so for any word w in the x_j and their inverses, $Aw \subseteq A$, that is $AG_n \subseteq A$. Now A contains $ey_0 = e$, and so

$$G_n = eG_n \subseteq AG_n \subseteq A \quad ,$$

and $A = G_n$, as claimed. Now in the subgroup H , all the relations (corresponding to R, S and T) of G_{n-1} hold (they are a subset of the corresponding relations in G_n), and so by the Substitution Test H is a homomorphic image of G_{n-1} , whence

$$|H| \leq |G_{n-1}| \leq (n-1)! \quad ,$$

by induction. Then,

$$|G_n| = |A| = |Hy_0 \cup \ldots \cup Hy_{n-1}| \leq n.|H| \leq n.(n-1)! = n! \quad ,$$

as required.

EXERCISE 1. Check the details of Step 4 in the proofs of Theorems 1 and 2.

EXERCISE 2. Use Tietze transformations to show that

$$<a,b,c \mid ab = c, \ bc = a, \ ca = b>$$

is a presentation of the quaternion group Q .

EXERCISE 3. Find a subgroup of $GL(2,C)$ isomorphic to D_n . Can you embed D_n in $GL(2,R)$?

EXERCISE 4. Let G be a finite group generated by two distinct elements of order 2 . Prove that G is dihedral.

EXERCISE 5. By finding a suitable concrete representation,

prove that the groups

$$\langle x,y \mid x^8 = y^2 = e, \; y^{-1}xy = x^3 \rangle$$

$$\langle x,y \mid x^8 = y^2 = e, \; y^{-1}xy = x^5 \rangle$$

both have order 16 . Can either of these groups be presented
with only two relations? Are they isomorphic?

EXERCISE 6. Describe the centre $Z(Q_{2n})$ of Q_{2n} , and use von
Dyck's theorem (3.3) to prove that $Q_{2n}/Z(Q_{2n}) \cong D_n$ for $n \geq 3$.
What happens when $n = 2$?

EXERCISE 7. Having observed that all the D_n are von Dyck
groups for $n \geq 3$, make a reasonable definition of D_2 .

EXERCISE 8. The presentation for S_4 given in Theorem 3 has 3
generators and 6 relations. Can you cut down on this? (cf.
Exercise 9).

EXERCISE 9. As a special case of Exercise 7, we have
$D_3 = D(2,2,3) = S_3$. S_4 is also a von Dyck group - which one?

EXERCISE 10. Prove that if $\xi,\eta,\xi\eta$ are distinct elements of
order 2 in $GL(2,C)$, then one of them is $-I$. Using the fact
that scalar matrices are central, deduce that no subgroup of
$GL(2,C)$ is isomorphic to S_n for $n \geq 4$.

EXERCISE 11. Prove that, for $n \in N$,

$$\langle x_1,\dots,x_n \mid x_i^3 \; (1 \leq i \leq n), \; (x_i x_j)^2 \; (1 \leq i < j \leq n) \rangle$$

is a presentation of the alternating group A_{n+2} .

§6. Finitely-generated abelian groups

We begin by elevating Exercises 3.4 and 4.3 to the status of theorems, for which purpose we fix the notation

$$X = \{x_1,\ldots,x_r\}, \quad C = \{[x_i,x_j] \mid 1 \leq i < j \leq r\}, \quad r \in N ,$$

and by san regard the set C of commutators as a subset of any group generated by X . Recall that the derived group (or commutator subgroup) G' of a group G is the subgroup of G generated by all commutators $[g_1,g_2]$, $g_1,g_2 \in G$, and that G' is a normal subgroup of G with abelian quotient group. G' is characterized by the fact that, for a normal subgroup N of G ,

$$G/N \text{ abelian} \iff G' \subseteq N , \tag{1}$$

and so we sometimes write $G/G' = G^{ab}$ (G abelianized).

Theorem 1. *If* $G = <X|R>$, *then* $G/G' = <X|R,C>$.

Proof. By von Dyck's theorem (3.3) on adjunction of relators, we merely have to show that G' coincides with normal closure $\bar{C}$ of C in G . Since the generators of $<X|R,C> = G/\bar{C}$ all commute, this group is abelian, and so $G' \subseteq \bar{C}$ by (1). On the other hand, G' is a normal subgroup of G containing C , whence $\bar{C} \subseteq G'$.

Theorem 2. *If* $F = F(X)$ *is free of rank* r , *then* F/F' *is:*

(i)   given by the presentation $<X|C>$,

(ii)   isomorphic to the direct product of r *copies of the infinite cyclic group,*

(iii) free abelian of rank r .

Proof. Part (i) is just the case $R = \emptyset$ of the previous theorem. To prove that the direct product $Z^{\times r}$ of r copies of Z has the presentation $<X|C>$, we proceed by induction on r , noting that C is empty when $r = 1$. For $r > 1$, assume that

51

$$Z^{\times(r-1)} = \langle x_1,\ldots,x_{r-1} \mid C_{r-1}\rangle, \quad C_{r-1} = \{[x_i,x_j] \mid 1 \le i < j \le r-1\} \; ,$$

whereupon

$$Z^{\times r} = Z^{\times(r-1)} \times Z = Z^{\times(r-1)} \times \langle x_r \mid \; \rangle \; , \; \text{say,}$$

$$= \langle X \mid C_{r-1}, [x_1,x_r],\ldots,[x_{r-1},x_r]\rangle \; , \; \text{by Theorem 4.2,}$$

and this is just $\langle X \mid C\rangle$.

Finally, to prove freeness, we assume that F^{ab} is the (internal) direct product of infinite cyclic groups generated by $x_1,\ldots,x_r$, in accordance with part (ii). Changing to additive notation, every element of F^{ab} is uniquely a Z-linear combination of $x_1,\ldots,x_r$. Now let G be any additively-written abelian group and $\theta\colon X \to G$ any mapping, and define $\theta'\colon F^{ab} \to G$ as follows. If $x \in F^{ab}$, we write $x = \sum_{i=1}^{r} k_i x_i$ $(k_i \in Z, \; 1 \le i \le r)$, and define

$$x\theta' = \sum_{i=1}^{r} k_i (x_i \theta) \; .$$

This clearly extends θ and is easily seen to be a homomorphism (cf. the proof of freeness of $F(X)$ in Theorem 1.2); it is unique since X generates F^{ab} .

Remark 1.   From now on, we denote the free abelian group on X by $A = A(X)$ and continue to write it additively. That its elements are unique Z-linear combinations of elements of X (as in the above proof) is just the abelian analogue of Theorem 1.3. Our next step towards the Basis Theorem consists of proving the analogues of Theorem 1.4 and 2.1 (Nielsen-Schreier).

Theorem 3.   If X generates an abelian group G , then there is an epimorphism $\theta\colon A(X) \to G$ fixing X elementwise. Every abelian group is a homomorphic image of some free abelian group.

Proof. To prove this, we can either give an abelian version of the proof of Theorem 1.4, or proceed as follows. If $G = \langle X|R \rangle$ is abelian, then $G = G/G' = \langle X|R,C \rangle$ and, shifting our weight on to the other foot as it were, we see (by von Dyck's theorem) that G is just the factor group of $A(X) = \langle X|C \rangle$ by $\bar{R}$.

Theorem 4 (Dedekind). *If $A = A(X)$ is free abelian of rank r and B is a subgroup of A , then B is free abelian of rank at most r .*

Proof. We go by induction on r , noting that the case $r = 1$ is just a well known result of elementary group theory. Now let $r > 1$ and assume the result for $r - 1$. Define subgroups

$$H = \langle x_1, \ldots, x_{r-1} \rangle , \quad Z = \langle x_r \rangle$$

of A , so that H is free abelian of rank $r - 1$ and $A = H \times Z$. By the inductive hypothesis, $B \cap H$ is free abelian - on $y_1, \ldots, y_s$ say, with $s \leq r - 1$. Furthermore,

$$\frac{B}{B \cap H} \cong \frac{B+H}{H} \leq \frac{A}{H} \cong Z ,$$

so that $B/B \cap H$ is either trivial or infinite cyclic (by the case $r = 1$). In the first case, $B = B \cap H$ and we are done, so we assume that $B/B \cap H = \langle b + B \cap H \rangle$ with $b \in B \backslash H$. We write $b = h + \ell x_r$, $h \in H$, $\ell \in \mathbb{N}$, and claim that B is free abelian on the set $Y = \{y_1, \ldots, y_s, b\}$. To do this, we invoke the abelian analogue of Theorem 1.3 (see Remark 1 above), noting first that Y clearly generates B . To prove $\mathbb{Z}$-linear independence of the set Y , suppose that

$$\sum_{i=1}^{s} k_i y_i + kb = 0 , \quad k_i, k \in \mathbb{Z} . \tag{2}$$

Thus we have

$$k\ell x_r = k(b-h) = -\sum_{i=1}^{s} k_i y_i - kh \in H \quad,$$

whence $k\ell x_r = 0$, since $H \cap <x>_r = \{0\}$. Since $\ell \neq 0$, we must
have $k = 0$, and (2) reduces to $\sum_{i=1}^{s} k_i y_i = 0$, and since the y_i
are *free* generators of $B \cap H$, each $k_i = 0$. Thus every element
of B is *uniquely* a Z-linear combination of the elements of Y ,
which proves our claim and hence the theorem.

In order to exhibit presentations for finitely generated
abelian groups and to describe the structure of derived factor
groups of arbitrary finitely presented groups, we need to recall
the Basis Theorem to the point of exhuming the bones of its proof,
upon which grisly proceeding we now bravely embark.

As above, let A be free abelian on $X = \{x_1,\ldots,x_r\}$ and
$U = \{u_1,\ldots,u_n\}$ another set of *free* generators of A . Corresponding to the relations $X = X(U)$, $U = U(X)$ in the proof of
Theorem 4.4, we have systems of equations

$$x_i = \sum_{j=1}^{n} p_{ij} u_j \quad, \quad 1 \leq i \leq r \quad, \tag{3}$$

$$u_j = \sum_{k=1}^{r} q_{jk} x_k \quad, \quad 1 \leq j \leq n \quad, \tag{4}$$

where the p_{ij}, q_{jk} are integers. Substituting (4) in (3),
uniqueness of expression yields that

$$\sum_{j=1}^{n} p_{ij} q_{jk} = \delta_{ik} \quad, \quad 1 \leq i,k \leq r \quad.$$

In other words, the integer matrices $P = (p_{ij})$, $Q = (q_{jk})$ have
the property that $PQ = I_r$. Similarly, the freeness of the
generators U guarantees that $QP = I_n$. This proves that $r = n$
and $Q = P^{-1}$. Conversely, any transformation of the type (4)
with $Q = (q_{ik})$ invertible over Z will yield a new set of free
generators of A .

Now let B be an arbitrary subgroup of A , so that by
Dedekind's theorem B is free abelian of rank s , with $s \leq r$.

Letting $Y = \{y_1, \ldots, y_s\}$ be a set of free generators of B , we have equations

$$y_k = \sum_{i=1}^{r} m_{ki} x_i , \quad 1 \le k \le s , \qquad (5)$$

and B is determined by the $s \times r$ matrix $M = (m_{ki})$, with respect to the sets Y, X of free generators. The effect on M of changing to generators U of A is found by substituting (3) into (5), which yields the matrix $MP = MQ^{-1}$. On the other hand, if Y is changed to another set V of free generators of B in accordance with an invertible $s \times s$ matrix T , then B is determined with respect to V, X by TM , and with respect to V, U by TMQ^{-1} .

In order to make this matrix as simple as possible we invoke the invariant factor theorem for integer matrices, which asserts that for any $s \times r$ integer matrix M , we can find $T \in GL(s, Z)$, $Q \in GL(r, Z)$ such that TMQ^{-1} is a diagonal matrix $D = \text{diag}(d_1, \ldots, d_\ell)$, with $\ell = \min(r, s)$ and where the d_i are non-negative integers each dividing its successor, and as such are uniquely determined by M . The d_i are called the *invariant factors* of M , and are constructed by a sequence of elementary row and column operations. For the uniqueness, we let $h_i(M)$ be the highest common factor of the i-rowed minors of M , $1 \le i \le m$, and observe that by basic (but nasty) linear algebra, these numbers are invariant under pre- and post-multiplication of M by invertible matrices. Thus, $h_i(M) = h_i(D)$ for all i , and a little thought shows that $h_i(D) = d_1 \ldots d_i$. Letting $h_0(M) = 1$ for convenience, we thus have $d_i = h_i(M)/h_{i-1}(M)$ for $1 \le i \le m$, showing that M determines the d_i .

Example 1. Let A be free on $\{x_1, x_2, x_3\}$ and let $B \le A$ be (freely) generated by

$$y_1 = 3x_1 - 3x_2 ,$$
$$y_2 = 2x_1 + 2x_2 + 2x_3 ,$$

so that

$$M = \begin{pmatrix} 3 & -3 & 0 \\ 2 & 2 & 2 \end{pmatrix} .$$

Transforming the x_i, y_i respectively by

$$Q = \begin{pmatrix} 1 & -5 & 2 \\ 0 & -2 & -1 \\ 0 & -1 & 0 \end{pmatrix} , \quad T = \begin{pmatrix} -1 & 1 \\ 2 & -3 \end{pmatrix} ,$$

we find that

$$D = TMQ^{-1} = \begin{pmatrix} 1 & 0 & 0 \\ 0 & 6 & 0 \end{pmatrix} ,$$

and indeed,

$$h_1(M) = hcf\{2, \pm 3, 0\} = 1 ,$$

$$h_2(M) = hcf\{-6, 6, 12\} = 6 .$$

Now let G be an abelian group, finitely generated by X say, and regard G as a homomorphic image of $A = A(X)$ via the homomorphism θ of Theorem 3. Letting $Ker\ \theta$ play the role of B in the above, we have

$$G = Im\ \theta \cong A/Ker\ \theta = A/B ,$$

so that with respect to suitable new sets of free generators, B is determined by a diagonal matrix D of the above type. To be specific, this means that G is just the free abelian group on $\{u_1, \ldots, u_r\}$ factored out by the free abelian group on $\{d_1 u_1, \ldots, d_s u_s\}$. Using a simple isomorphism theorem for direct products (see Exercise 2), we see that G is just the direct product of the groups

$$Zu_i / Zd_i u_i \cong Z_{d_i} , \quad 1 \le i \le s ,$$

and $r-s$ infinite cyclic groups.

56

Theorem 5 (The Basis Theorem). *Given a finitely generated abelian group* G *, there are integers* $s, n \geq 0$ *and integers* $d_i \geq 2$ *, $1 \leq i \leq s$, each dividing its successor, such that*

$$G \cong Z_{d_1} \times \ldots \times Z_{d_s} \times \underbrace{Z \times \ldots \times Z}_{n \text{ copies}} .$$

Further, s, n *and the* d_i *are all determined by* G *.*

With the above notation, $n = r-s$ and those $d_i = 1$ have been ignored. The d_i are called the *invariant factors*, and n the *rank*, of G . Combining this with Theorem 4.2, we obtain presentations for all finitely generated abelian groups.

Theorem 6. *Every finitely generated abelian group* G *has one presentation of the form*

$$G = <X|P,C> \quad,$$

where

$$X = \{x_1, \ldots, x_{s+n}\} \quad,$$
$$P = \{x_i^{d_i} \mid 1 \leq i \leq s\} \quad,$$
$$C = \{[x_i, x_j] \mid 1 \leq i < j \leq s+n\} \quad,$$

the d_i *being integers* ≥ 2 *, each dividing its successor.*

There now arises the natural question as to how we can compute the rank and invariant factors of G/G' when G is a group specified by some finite presentation $<X|R>$. The answer is encapsulated in the above discussion and runs as follows. We have

$$G/G' = <X|R,C> = A(X)/\bar{R} \quad,$$

so we take $A = A(X)$, $B = \bar{R} = <R>$ (as A is abelian). If

$X = \{x_j \mid 1 \le j \le r\}$ and $R = \{w_i \mid 1 \le i \le s\}$, we just write the w_i in additive notation:

$$w_i = \sum_{j=1}^{s} m_{ij} x_j , \quad 1 \le i \le s ,$$

where m_{ij} is the *exponent-sum* of x_j in w_i , and $M = (m_{ij})$ is called the *relation matrix* of the presentation $<X|R>$. The only difference between this M and that used above to determine the subgroup $B = <R>$ of A , is that the generators of B corresponding to the rows of M may not be free. Nevertheless, free generators do exist and can be obtained by performing elementary row operations on M . This new matrix and M thus have the same canonical diagonal form $D = \mathrm{diag}\{d_1,\ldots,d_\ell\}$, with $\ell = \min(r,s)$. Note that the divisibility property of the d_i's means that any 1's occur at the beginning and any 0's at the end. The invariant factors of G/G' are just the remaining d_i's , and its rank is r minus the number of non-zero d_i's .

Example 2. We compute the structure of G/G' when

$$G = <x,y,z \mid (xyz)^2 = e, \ x^3 = y^3, \ (zxy)^4 = e> .$$

The relation matrix here is

$$M = \begin{pmatrix} 2 & 2 & 2 \\ 3 & -3 & 0 \\ 4 & 4 & 4 \end{pmatrix} \sim \begin{pmatrix} 3 & -3 & 0 \\ 2 & 2 & 2 \\ 0 & 0 & 0 \end{pmatrix}$$

via elementary row operations. Using Example 1 above, the invariant factors of M are 1,6,0 and so G/G' has one invariant factor, namely 6 , and rank 3-2 = 1 . Hence, $G/G' = Z_6 \times Z$.

Example 3. The presentation

$$G = <x,y,z \mid x^y = x^{-1}y^4, \ y^z = y^{-1}z^4, \ z^x = z^{-1}x^4>$$

has relation matrix

$$M = \begin{pmatrix} 2 & -4 & 0 \\ 0 & 2 & -4 \\ -4 & 0 & 2 \end{pmatrix} .$$

We clearly have $h_1(M) = 2$, $h_2(M) = 4$ and $h_3(M) = -\det M = 56$. Hence,

$$G/G' = Z_2 \times Z_2 \times Z_{14} .$$

We conclude this chapter and pave the way for the next with the following intuitively reasonable result.

Theorem 7.   *If $G = \langle X | R \rangle$ is a finite presentation of a finite group G, then $|X| \leq |R|$.*

Proof.   To prove the contrapositive assertion, assume that $|X| > |R|$, so that the relation matrix has fewer rows than columns. It follows that G/G' has fewer invariant factors than generators and thus is infinite.

EXERCISE 1.   Let F be a free group of rank r. Use diagrams to give a direct proof that F/F' is free abelian of rank r.

EXERCISE 2.   Let H_i, G_i be groups with $H_i \lhd G_i$, $i = 1,2$. Prove that $H_1 \times H_2 \lhd G_1 \times G_2$ and that

$$\frac{G_1 \times G_2}{H_1 \times H_2} \cong \frac{G_1}{H_1} \times \frac{G_2}{H_2} .$$

Generalize this to a direct product of n groups, $n \in N$.

EXERCISE 3.   Let $F = \langle X | \ \rangle$ be a free group of finite rank n and $G = \langle X | R \rangle$ a group of finite order g. Let U be a (right) transversal for $H = \bar{R}$ in F. Prove that the $g.|R|$ cosets

$$\{H'u^{-1}ru \mid u \in U, \ r \in R\}$$

generate H/H' , and use the Nielsen-Schreier theorem to derive
an alternative proof of Theorem 7.

EXERCISE 4.   Write down a relation matrix for the standard
presentation of S_4 (see Theorem 5.3) and use it to compute S_4^{ab} .

EXERCISE 5.   Let A be the (additively-written) free abelian
group in $\{x,y,z\}$, and B the subgroup of A generated by

$$\{21x + 18y + 15z, 9x + 6y + 15z, 12x + 18y + 6z\} \quad ;$$

compute the invariant factors of A/B .

EXERCISE 6.
(i)   Determine the structure of G/G' when G is the group

$$\langle x,y \mid y^{-1}xy = x^r \rangle , \quad r \in \mathbb{N} .$$

(ii)   Compute the order of G/G' when G is given by

$$\langle x,y,z \mid x^y = y^{b-2}x^{-1}y^{b+2}, y^z = z^{c-2}y^{-1}z^{c+2}, z^x = x^{a-2}z^{-1}x^{a+2} \rangle ,$$

where a,b,c are non-zero even integers.
(iii)   Determine the structure of G/G' when G is the Fibonacci
group

$$F(2,n) = \langle x_1,\ldots,x_n \mid x_1x_2 = x_3, x_2x_3 = x_4,\ldots,x_nx_1 = x_2 \rangle$$

for n = 5,6 . [Hint: Use Tietze transformations to reduce the
number of generators.]
(iv)   Find a formula for the order of the derived factor group of
F(2,n) in terms of the Lucas numbers

$$1,3,4,7,11,18,\ldots \quad .$$

(v) Describe G/G' when G is the von Dyck group $D(\ell,m,n)$.

(vi) Let him that hath understanding find the order of the group

$$G = \langle x,y,z,t \mid x^3y^7 = y^4x^7 = z^3t^5 = t^4z^6 = [x,z] = e \rangle \quad .$$

[Hint: It is the number of a man.]

3 · Groups with few relations

Fate chooses your relations, you choose your
friends (Delille: Malheur et pitié)

We have just shown (Theorem 6.7) that for a finite group
$G = <X|R>$, we must have $|X| \leq |R|$, whence finite groups have
non-positive deficiency, in the following sense.

Definition. We define the *deficiency* of a finitely presented
group G by

$$\text{def } G = \max\{|X| - |R| \mid \text{all finite presentations } <X|R> \text{ of } G\} \ .$$

Definition. For a *finite* group $G = <X|R>$, the (Schur) multipli-
cator of G is defined by

$$M(G) = \frac{F' \cap \bar{R}}{[F, \bar{R}]} \ ,$$

where $\bar{R}$ is the normal closure of R in $F = F(X)$.

Among other things, Schur proved in 1907 that $M(G)$ is
(i)   an *invariant* of G , i.e. independent of the finite presen-
       tation $<X|R>$,
(ii)   a *finite* abelian group,
(iii) generated by $-\text{def } G$ elements.

Though we shall shed some light on these and other results
later, their proofs are beyond our present scope. For the present,
we concern ourselves with finite groups of deficiency zero, noting
that these must have trivial multiplicator. The class of finite
groups with trivial multiplicator but non-zero deficiency was
shown by Swan to include some soluble groups; whether it contains
any nilpotent groups is an unsolved problem. We content ourselves

62

here with the problem of exhibiting certain finite groups of deficiency zero, hereafter referred to as *interesting* groups.

§7. Metacyclic groups

We have so far encountered only two types of interesting groups, namely the cyclic groups Z_n (Theorem 3.2(ii)) and the quaternionic groups Q_{2n} (Theorem 5.1). The latter are sometimes referred to as dicyclic groups and are special cases of metacyclic groups. These are best understood in the context of Chapter VI as group extensions, of which they comprise an important special case.

Definition. A group G is called *metacyclic* if it has a normal subgroup H such that both H and G/H are cyclic.

In the case where G is finite, we can thus assume that

$$H = <x> \cong Z_m, \quad G/H = <Hy> \cong Z_n , \tag{1}$$

with $x,y \in G$, $m,n \in N$. Since H is a normal subgroup of index n , both $y^{-1}xy$ and y^n must belong to H , say

$$y^{-1}xy = x^r , \quad y^n = x^s , \tag{2}$$

where r,s are integers with $1 \leq r, s \leq m$. Now it follows from the first relation of (2) that for $a,b \in Z$, $b \geq 0$,

$$y^{-b}x^ay^b = x^{ar^b} , \tag{3}$$

(see Exercise 1), so that the second relation of (2) implies that

$$x = x^{-s}xx^s = y^{-n}xy^n = x^{r^n} ,$$

whence, $r^n \equiv 1 \pmod{m}$, since $|x| = |H| = m$. Similarly, we have

$$x^s = y^n = y^{-1}y^ny = y^{-1}x^sy = x^{rs} ,$$

so that $rs \equiv s \pmod{m}$. We are now in a position to write down presentations for all finite metacyclic groups.

Theorem 1. *Consider the group*

$$G = \langle x,y \mid x^m = e , y^{-1}xy = x^r, y^n = x^s \rangle , \qquad (4)$$

where

$$m,n,r,s \in N , \quad r,s \leq m ,$$

and

$$r^n \equiv 1 , \quad rs \equiv s \pmod{m} . \qquad (5)$$

Then $N = \langle x \rangle$ *is a normal subgroup of* G *such that*

$$N \cong Z_m , \quad G/N \cong Z_n .$$

Thus, G *is a finite metacyclic group, and moreover, every finite metacyclic group has a presentation of this form.*

Proof. To prove that $N \triangleleft G$, write any member w of G as a word in x,y . Since $x^m = e$ and $y^n = x^s$, we can write w as a *positive* word and consider it as an alternating product of syllables of the form x^i, y^j with $i,j \in N$. Now conjugation of any power of x by either of these again yields a power of x , by the consequence (3) of the relation $y^{-1}xy = x^r$. Hence $N^w \subseteq N$ for all $w \in G$ and $N \triangleleft G$. It now follows from Theorem 3.3 that G/N is given by adjoining the relation $x = e$ to the presentation (4), and this plainly yields Z_n .

While we know already that N is a factor group of Z_m , it is a non-trivial matter to prove that these groups are actually isomorphic. To do this, consider the set C of ordered pairs (i,j) where i is an integer with $0 \leq i \leq n-1$ and j is a

residue class modulo m . We define a binary operation on C by
setting

$$(i,j)(k,\ell) = \begin{cases} (i+k, \; \ell+jr^k) & , \;\; \text{if} \;\; i+k < n \;, \\ (i+k-n, \; \ell+jr^k+s) & , \;\; \text{if} \;\; i+k \geq n \;, \end{cases} \tag{6}$$

where r and s are as in the statement of the theorem. We
claim that C is a group, first observing that (0,0) is an
identity, $(0,j)^{-1} = (0,-j)$ and for $i > 0$, it follows from (5)
that

$$(i,j)^{-1} = (n-i,-jr^{n-i}-s) \;\; .$$

Checking the associative law is rather tedious, and we merely observe that because of (5), the product of $(i,j),(k,\ell),(a,b)$ is
equal to

$$(i+k+a, b+\ell r^a+jr^{k+a}) \;,$$
$$(i+k+a-n, b+\ell r^a+jr^{k+a}+s) \;,$$
$$(i+k+a-2n, b+\ell r^a+jr^{k+a}+2s) \;,$$

according as $[(i+k+a)/n] = 0,1,2$ respectively, and is thus independent of the bracketing. Now in C ,

$$(i,\ell) = (i,0)(0,\ell) = (1,0)^i(0,1)^\ell \;,$$

so that (1,0) and (0,1) generate C . Substitution of these
for y and x respectively in the relations of (4) yields
identities, so that C is a factor group of G by the
Substitution Test. Hence C is a concrete realization of G ,
and $|N| = m$ as required. The last part of the theorem has
already been proved above.

While this theorem gives a description of any finite metacyclic
group in terms of the four parameters m,n,r,s , it is *not* a
classification theorem. In fact, the problem of when two groups
of the type (4) are isomorphic is unsolved in general, though the

special case when mn is a prime-power has recently been dealt with by F.R. Beyl.

We see from the presentation (4) that def G is either -1 or 0 and so M(G) must be cyclic by the classical results of Schur mentioned in the introduction to this chapter, and by the same token, M(G) is trivial if def G = 0 . That the converse holds for finite metacyclic groups was shown by J.W. Wamsley in 1970, who gave a 2-generator 2-relation presentation when M(G) = E . A slicker version was given 3 years later by Beyl, and we describe this now.

There are two problems involved:
(i)   to compute $|M(G)|$ in terms of m,n,r,s ,   and
(ii) to define   G   by two relations when $M(G) = E$.
We shall give a paraphrase of (i) and prove (ii) in full (Theorem 2 below). To get at M(G) , we need to take a closer look at the congruences (5). From the first of these, it follows that

$$h = \frac{1}{m} (m,r-1)(m,1+r+\ldots+r^{n-1})$$

is an integer. From the second congruence, there is an integer k such that

$$s = \frac{km}{r-1} = \frac{k.m/(m,r-1)}{(r-1)/(m,r-1)} = \ell m/(m,r-1) \quad ,$$

where ℓ is an integer since $m/(m,r-1)$ and $(r-1)/(m,r-1)$ are coprime. By changing generators and using elementary number theory, ℓ can be replaced by (ℓ,h) , so that we can take $s = \ell m/(m,r-1)$ in (4), with ℓ a divisor of   h . Beyl's main result now asserts that ℓ is nothing other than $|M(G)|$. Granting this, the deficiency problem for finite metacyclic groups is solved as follows.

Theorem 2.   *Let*

$$G = <x,y \mid x^m = e , y^{-1}xy = x^r, y^n = x^s> \quad ,$$

66

where $m, n, r, s \in \mathbb{N}$ *and*

$$r^n \equiv 1 \pmod{m} \quad \text{and} \quad s = m/(m, r-1) \quad .$$

Then G *is presented, in terms of the same generators* x *and* y , *by*

$$<x,y \mid y^n = x^s, \ [y, x^{-t}] = x^{(m, r-1)}> \quad , \tag{7}$$

where t *is a certain integer.*

Proof.

(i) We first construct an integer t such that the second relation of (7) holds in G . To this end, let

$$(m, r-1) = u(r-1) + vm \quad ,$$

so that $(u, s) = 1$. Now let w be the largest factor of m coprime to u , and put $t = u + ws$. Now modulo m ,

$$(m, r-1) \equiv u(r-1) \equiv t(r-1) \quad ,$$

since $s(r-1)$ is divisible by m . Hence, in G ,

$$[y, x^{-t}] = (y^{-1} x^t y) x^{-t} = x^{(r-1)t} = x^{(m, r-1)} \quad ,$$

and the relations of (7) hold in G .

(ii) We now assume the relations of (7) and show that they define G . Armed with the knowledge that x^s commutes with y and $[y, x^{-t}]$ commutes with x , we see that

$$x^{ts} = y^{-1} x^{ts} y = (y^{-1} x^t y)^s = ([y, x^{-t}] x^t)^s = [y, x^{-t}]^s x^{ts} \quad ,$$

and so

$$x^m = x^{(m, r-1)s} = [y, x^{-t}]^s = e \quad .$$

To complete the proof, we first need to show that t and m are coprime, so assume (for a contradiction) that p is a prime factor of both. Since p divides m, it divides exactly one of w and u. But p divides $t = u + ws$, so it cannot divide w. Hence, p divides u and s contrary to the fact that $(u,s) = 1$. Thus, there is an integer k such that $kt \equiv 1 \pmod{m}$, and we have

$$[y,x^{-1}] = [y,x^{-kt}] = ([y,x^{-t}]x^t)^k x^{-kt} = [y,x^{-t}]^k$$

$$= x^{k(m,r-1)} = x^{kt(r-1)} = x^{r-1}$$

Hence $x^y = x^r$, and the three relations of G follow from the two relations of (7).

EXERCISE 1. Let x and y be members of a group G such that

$$y^{-1}xy = x^r, \quad r \in \mathbb{Z}.$$

Prove that for $a,b \in \mathbb{Z}$ with $b \geq 0$,

$$y^{-b}x^a y^b = x^{ar^b}.$$

EXERCISE 2. (Cf. Exercise 6.6(i).) Consider the 'archetypal metacyclic group'

$$G = \langle x,y \mid y^{-1}xy = x^r \rangle .$$

Prove that every element in G can be written in the form

$$y^i x^j y^{-k}, \quad i,j,k \in \mathbb{Z}, \ i,k \geq 0.$$

Identify the elements of G', and deduce that G' is abelian. Do you believe that G really is metacyclic?

EXERCISE 3. A group G is called *split metacyclic* if it has a cyclic normal subgroup with a cyclic complement. Such a group thus has a presentation of the form

$$G = <x,y \mid x^m = e , \quad x^y = x^r , \quad y^n = x^m> \quad .$$

Use Beyl's criterion to prove that $M(G)$ is cyclic of order

$$(m,r-1)(m,1+r+\ldots+r^{n-1})/m \quad .$$

EXERCISE 4. Use the result of the previous exercise to compute the multiplicator of the dihedral group D_m of degree m . Write down a 2-generator 2-relation presentation of D_m when m is odd.

EXERCISE 5. Compute $M(Z_m \times Z_n)$, $m,n \in N$.

EXERCISE 6. Let p be an odd prime, $a,b \in N$ and $k \in Z$ such that $(k,p) = 1$. Prove that

$$(1+kp^a)^{p^b} = 1 + kp^{a+b} + \ell p^{a+b+1} \quad ,$$

with $\ell \in Z$.

EXERCISE 7. Use Exercise 6 to find a deficiency-zero presentation of the group

$$G = <x,y \mid x^{p^{a+b}} = e, \quad y^{-1}xy = x^{1+p^a} , \quad y^{p^b} = e>$$

when p is an odd prime and $a,b \in N$.

EXERCISE 8. Prove that the groups G of Exercise 7 exhaust the class of split metacyclic groups with odd prime-power order and trivial multiplicator.

EXERCISE 9. Prove that the group

$$<x,y \mid x^m = y^n = [x,y]>$$

is cyclic when $(m,n) = 1$.

§8. Interesting groups with three generators

Twenty years ago, the list of known interesting groups comprised

(i)   cyclic groups,

(ii)   certain metacyclic groups (see §7) studied by Schur,

(iii) certain derivatives of the von Dyck groups (see Example 4.1)
      studied by Miller.

Now all these groups are 2-generated, and the question naturally arose as to whether there were any interesting groups needing 3 generators or more. The first examples were provided by J. Mennicke in 1959, who showed that the groups

$$M(a,b,c) = \langle x,y,z \mid y^{-1}xy = x^a,\ z^{-1}yz = y^b,\ x^{-1}zx = z^c \rangle$$

are finite in the case $a = b = c \geq 3$. Eleven years later, Wamsley showed that the two classes of groups

$$W_\pm(a,b,c) = \langle x,y,z \mid x^z = x^a,\ y^{z^{\pm 1}} = y^b,\ z^c = [x,y] \rangle$$

were interesting provided $(a-1)(b-1)c \neq 0$, this work being based on the groups $\text{Mac}(a,b) = W_-(a,b,1)$ discovered by I.D. Macdonald in 1962. While our current stock of interesting groups with 2-generators is very large, the list of those needing 3 generators is fairly short. It consists of the 3-generator groups in the above three classes of Mennicke and Wamsley, together with the

$$J(a,b,c) = \langle x,y,z \mid x^y = y^{b-2}x^{-1}y^{b+2},\ y^z = z^{c-2}y^{-1}z^{c+2},$$
$$z^x = x^{a-2}z^{-1}x^{a+2} \rangle \ ,$$

where a,b,c are non-zero *even* integers (see Exercise 6.6(ii)). Since the $J(a,b,c)$ are the easiest to analyse we investigate this class now, observing that similar methods apply to the other three classes.

Let $G = J(a,b,c)$. Our first step is to conjugate the first relation by y :

$$x^{y^2} = y^{b-2}(x^y)^{-1}y^{b+2} = y^{-4}xy^4 = x^{y^4} ,$$

so that x commutes with y^2 (write $x \sim y^2$). From the other two relations, we have

$$y \sim z^2, \ z \sim x^2 , \tag{1}$$

and we deduce that the subgroup $H = \langle x^2, y^2, z^2 \rangle$ of G is *abelian*. Furthermore, we deduce from the original relations that

$$x^y = x^{-1}y^{2b}, \ y^z = y^{-1}z^{2c}, \ z^x = z^{-1}x^{2a} , \tag{2}$$

and as a consequence,

$$(x^2)^x = x^2, \ (x^2)^y = x^{-2}y^{4b}, \ (x^2)^z = x^2$$

all belong to H . Since the conjugates of x^2 by x^{-1}, y^{-1}, z^{-1} are equal to its conjugates by x, y, z respectively (H is abelian), we see that $(x^2)^w \in H$ for any $w \in G$. Similarly, $(y^2)^w, (z^2)^w \in H$ and we deduce that H is *normal* in G . By Theorem 3.3, a presentation for G/H is given by adjoining the relations $x^2 = y^2 = z^2 = e$ to those defining G , and after a couple of Tietze transformations, we have

$$G/H = \langle x, y, z \mid x^2 = y^2 = z^2 = (xy)^2 = (yz)^2 = (zx)^2 = e \rangle .$$

Thus, $G/H \cong Z_2 \times Z_2 \times Z_2$ and $|G:H| = 8$.

We now apply the relations (1) and (2) together with the fact that $x^2 \sim y^2$ to compute that

$$[[x,y],z^x] = [x^{-2}y^{2b}, z^{-1}x^{2a}] = [y^{2b}, z^{-1}] = [y^{2b}, z] = y^{-2b}(y^z)^{2b}$$
$$= y^{-2b}(y^{-1}z^{2c})^{2b} = y^{-4b}z^{4bc} .$$

Substituting this and its companions into the Witt identity (see

Exercise 4.10), we have:

$$e = [[x,y],z^x][[z,x],y^z][[y,z],x^y]$$

$$= y^{-4b}z^{4bc} \cdot x^{-4a}y^{4ab} \cdot z^{-4c}x^{4ca}$$

$$= x^{4a(c-1)}y^{4b(a-1)}z^{4c(b-1)} ,$$

since x^2, y^2, z^2 all commute. Since y^2, z^2 commute with y so does $x^{4a(c-1)}$, and we have

$$x^{4a(c-1)} = (x^{4a(c-1)})^y = (x^{-1}y^{2b})^{4a(c-1)} = x^{-4a(c-1)}y^{8ab(c-1)} .$$

Hence

$$x^{8a(c-1)} = y^{8ab(c-1)} ,$$

and similarly,

$$y^{8b(a-1)} = z^{8bc(a-1)} , \quad z^{8c(b-1)} = x^{8ca(b-1)} .$$

Using each of these three relations, we finally obtain:

$$x^{8a(c-1)(a-1)(b-1)} = y^{8ab(c-1)(a-1)(b-1)}$$

$$= z^{8abc(c-1)(a-1)(b-1)}$$

$$= x^{8a^2bc(c-1)(a-1)(b-1)} ,$$

showing that x has order dividing

$$|8a(c-1)(a-1)(b-1)(abc-1)| , \tag{3}$$

which is non-zero. Thus x^2 has finite order and similarly, so do y^2 and z^2 . Since x^2, y^2, z^2 all commute, it follows that $H = \langle x^2, y^2, z^2 \rangle$ is finite. Since $G/H = Z_2 \times Z_2 \times Z_2$ needs 3 generators, so does G , and we can state the following theorem.

Theorem 1. *Let* G *be the group*

$$\langle x,y,z \mid x^y = y^{b-2}x^{-1}y^{b+2}, \ y^z = z^{c-2}y^{-1}z^{c+2}, \ z^x = x^{a-2}z^{-1}x^{a+2}\rangle \ ,$$

where a,b,c *are non-zero even integers, and let*
$H = \langle x^2, y^2, z^2 \rangle \leq G$. *Then* H *is an abelian normal subgroup of*
G *with* G/H *elementary abelian of order* 8 . *The order of*
G *is a divisor of*

$$512|abc|.|(a-1)(b-1)(c-1)(abc-1)|^3 \ . \tag{4}$$

The bound for $|G|$ is obtained by multiplying by 8 the
product of the bounds for $|x^2|, |y^2|, |z^2|$ obtained from (3) and
its analogues. Even in the simplest case, namely when
a = b = c = 2 , this yields a 7-digit number, while the order of
G in this case is 7.2^{11} (the bound obtained by Wamsley). We
must defer until the next chapter the method for computing $|G|$
exactly (see Exercise 12.10), merely observing that the correct
value is

$$|G| = 2^8|abc(abc-1)| \ . \tag{5}$$

We have proved that G is a *metabelian* group, that is, a group
with abelian derived group. Recall that a group is called *nil-potent* if it has a finite chain of normal subgroups

$$G = G_1 > G_2 > \ldots > G_{n+1} = E$$

with $G_i/G_{i+1} \subseteq Z(G/G_{i+1})$ for all i . The *class* of a nilpotent
group is the least value of n for which such a chain exists, so
that for example, nilpotent groups of class 1 are just abelian
groups. Simple commutator calculations show that J(a,b,c) is
nilpotent if and only if $|abc|$ is a 2-power, whereupon its
class is equal to $3 + \log_2 \max\{|a|, |b|, |c|\}$.

Whether or not there is an interesting group that needs 4
generators is an unsolved problem; the best we can do is 4 gener-

ators and 6 relations. That all interesting *nilpotent* groups are 3-generated follows from the Golod–Šafarevič theorem (see §21), but very little is known about the general case.

EXERCISE 1.   Use (5) to compute the order of $J(a,b,c)'$ (see Exercise 6.6(ii)).

EXERCISE 2.   By studying relations derived in the proof of Theorem 1, try to reduce as far as possible the bound (4) for the order of $J(a,b,c)$.

EXERCISE 3.   When do abelianized Mennicke and Wamsley groups need 3 generators? Write down necessary and sufficient conditions on a,b,c in each case.

EXERCISE 4.   Identify $M(a,a,a)$ when $a = 0,1,2$.

EXERCISE 5.   Use the Witt identity to prove that x,y,z each have finite order in $M(a,b,c)$ when $|a-1|,|b-1|,|c-1|$ are all at least 2.

EXERCISE 6.   Prove that every element of $M(a,b,c)$ can be written in the form $x^i y^j z^k$, $i,j,k \in Z$. Deduce that $M(a,b,c)$ is finite when $|a-1|,|b-1|,|c-1|$ are all at least 2, and write down a bound for the order in this case.

EXERCISE 7.   Find a finite group that needs 4 generators and can be defined by 6 relations.

§9. Cyclically presented groups

Cyclically presented groups comprise a potentially rich source of interesting groups, and indeed we have already looked at several examples of this type of group. We now give a formal definition, though the name is self-explanatory.

Definition 1.   Let $F = \langle x_1, \ldots, x_n | \rangle$ and let θ be the auto–

morphism of F induced by permuting the subscripts of the free
generators in accordance with the cycle $(1 \ 2 \ \ldots \ n) \in S_n$. For
any reduced word $w \in F$, the *cyclically presented group* $G_n(w)$
is given by

$$G_n(w) = \langle x_1, x_2, \ldots, x_n \mid w, w\theta, \ldots, w\theta^{n-1} \rangle \quad .$$

Since cyclically presented groups have non-negative deficiency,
$G_n(w)$ is interesting if and only if it is finite. Examples of
such groups appearing in the previous section are

$$M(a,a,a) = G_3(x_2^{-1} x_1 x_2 x_1^{-a}) \ ,$$

$$J(a,a,a) = G_3(x_1^{-1} x_2^{a-1} x_1^{-1} x_2^{a+1}) \ ,$$

$$Mac(a,a) = G_2(x_1^{[x_1,x_2]} x_1^{-a}) \quad .$$

Further examples are provided by the Fibonacci groups
$F(2,n) = G_n(x_1 x_2 x_3^{-1})$ of Exercise 6.6(iii), and these are a
special case of the following class.

Definition 2. For $r,n \in N$ with $r \geq 2$, the *Fibonacci group*
$F(r,n)$ is defined as the cyclically presented group

$$F(r,n) = G_n(x_1, \ldots, x_r x_{r+1}^{-1}) \ ,$$

where subscripts are understood to be reduced modulo n to lie
in the set $\{1,2,\ldots,n\}$.

The following table gives an idea of what the $F(r,n)$ look
like for small values of r and n . In the (r,n) place is
written the order of $F(r,n)$, or the isomorphism type where
appropriate. The gaps in the table correspond to gaps in our
knowledge. We have already computed some of these entries and
others figure among the next set of exercises. The hardest to
identify is probably Campbell's group $F(3,6)$ - one of a
bewildering array of cyclically presented groups of order 1512
studied by C.M. Campbell and E.F. Robertson - and this is the

only finite entry in the table which is not metacyclic.

r \ n	1	2	3	4	5	6
2	E	E	Q	Z_5	Z_{11}	∞
3	Z_2	Q	Z_2	∞	Z_{22}	1512
4	Z_3	Z_3	63	Z_3	∞	
5	Z_4	24	∞	624	Z_4	∞
6	Z_5	Z_5	Z_5	125	7775	Z_5
7	Z_6	48	342	∞		117648

It is known that for fixed r , the $F(r,n)$ are eventually infinite; it is a nice exercise in small cancellation theory (see Chapter VII) to prove that $F(r,n)$ is infinite for $n > 5r$. The case $r = 2$ is thus almost completely decided. A machine implementation of Todd-Coxeter coset enumeration (see Chapter IV) shows that $F(2,7) \cong Z_{29}$, while A.M. Brunner has proved that $F(2,8)$ and $F(2,10)$ are both infinite (see Exercise 10 and Exercises 20.16-19). This leaves one group, namely $F(2,9)$, and the best we can say about this is as follows. By using a formidable array of algorithms and a considerable amount of ingenuity, G. Havas, J.S. Richardson, L.S. Sterling, a Univac 100/42 and a DEC KA 10 have proved that $F(2,9)$ has order at least 152.5^{741} .

We now turn to the study of the $G_n(w)$ in general, and concentrate on describing the structure of $A_n(w) = G_n(w)^{ab}$ by means of the theory in §6. In particular, we shall find necessary and sufficient conditions for $A_n(w)$ to be (i) infinite, (ii) perfect, and to do this we need the following definition.

Definition 3. The *polynomial* $f(t) = f_{n,w}(t)$ *associated with* the cyclically presented group $G = G_n(w)$ is given by

$$f(t) = \sum_{i=1}^{n} a_i t^{i-1} , \tag{1}$$

where a_i is the exponent sum of x_i in w , $1 \leq i \leq n$.
Since the n permutants of w under powers of θ (see

Definition 1) comprise a set of defining relators for $G_n(w)$, it follows that the matrix

$$
C = \begin{pmatrix}
a_1 & a_2 & a_3 & \cdots & a_{n-1} & a_n \\
a_n & a_1 & a_2 & \cdots & a_{n-2} & a_{n-1} \\
\cdots & & \cdots & & \cdots & \\
a_2 & a_3 & a_4 & \cdots & a_n & a_1
\end{pmatrix}
\tag{2}
$$

is a relation matrix for $A_n(w)$. This is a *circulant* matrix and its determinant is known.

Theorem 1. *With the notation of (1) and (2)*

$$
\det C = \prod_{i=1}^{n} f(\omega_i) \quad ,
$$

where ω_i ranges over the set of complex nth roots of unity.

Proof. Let ω be a primitive nth root of unity, and let $V = (v_{ij})$ be the Vandermonde matrix given by $v_{ij} = \omega^{ij}$. Now the (i,j) entry of the product CV is equal to $\omega^{ij} f(\omega^j)$, whence

$$
\det(CV) = [\prod_{j=1}^{n} f(\omega^j)] \det V \quad .
\tag{3}
$$

But the (i,k) entry of V^2 is equal to

$$
\sum_{j=1}^{n} \omega^{(i+k)j} = \begin{cases} 0 \, , & \text{if } i+k \neq n,2n \, , \\ n \, , & \text{if } i+k = n,2n \, , \end{cases}
$$

so that V^2 is just n times a permutation matrix. In particular, V is non-singular, so the theorem follows from (3).

In accordance with the theory of §6, we now have a formula for the order of $A_n(w)$.

Theorem 2. *If f is the polynomial associated with w , then*

$$
|A_n(w)| = \pm \prod_{\xi^n=1} f(\xi) \quad .
\tag{4}
$$

Note that the sign in (4) is chosen so as to make the right-hand side positive, and that we observe the convention usual in this context that $0 = \infty$.

Theorem 3. *The following three assertions are equivalent:*
(a) $A_n(w)$ *is infinite,*
(b) $f(t)$ *has a root in common with* $t^n - 1$,
(c) $f(t)$ *is a zero-divisor in the ring* $Z[t]/(t^n-1)$.

Proof. The equivalence of (a) and (b) follows at once from (4). Assume (b) and let ω be a common root. Suppose ω is a primitive kth root of unity $(1 \le k \le n)$ and let $\phi_k(t)$ be the kth cyclotomic polynomial. Since $f(t)$ and $t^n - 1$ are both integral polynomials, they are both divisible by $\phi_k(t)$:

$$t^n - 1 = \phi_k(t)p(t) , \quad f(t) = \phi_k(t)q(t)$$

say, where $p(t), q(t) \in Z[t]$. It follows that $f(t)p(t)$ belongs to the ideal (t^n-1) , and since the degree of $p(t)$ is less than n , (c) follows. Conversely, let (c) hold so that there are polynomials $p(t), g(t) \in Z[t]$ such that

$$f(t)p(t) = (t^n-1)g(t) ,$$

with $p(t)$ not divisible by $t^n - 1$. Now the factorization

$$t^n - 1 = \prod_{k|n} \phi_k(t)$$

into irreducibles shows that for some k , $\phi_k(t)$ is not a divisor of $p(t)$. By uniqueness of factorization, $\phi_k(t)$ is a divisor of $f(t)$, and assertion (b) follows.

Theorem 4. $A_n(w)$ *is trivial if and only if* $f(t)$ *is a unit in the ring* $Z[t]/(t^n-1)$.

Proof. Suppose first that $f(t)$ is a unit, and let $g(t) \in Z[t]$

be such that

$$f(t)g(t) \equiv 1 \ (mod(t^n-1)) \ .$$ (5)

We can of course assume that $g(t)$ has degree at most $n-1$, and it is a consequence of Theorem 1 that any such polynomial has the property

$$\prod_{\xi^n=1} g(\xi) \in Z \ .$$ (6)

That $A_n(w)$ is trivial now follows from (4), (5) and (6). To prove the converse let $F = \langle x_1, \ldots, x_n | \ \rangle$ and consider the mapping

$$\pi: F \to Z[t]/(t^n-1)$$
$$u \mapsto h(t) + (t^n-1) \ ,$$

where h is the polynomial associated with the word u. It is clear that π is an epimorphism of groups, and that Ker π, being the set of words with all exponent-sums zero, is just F'. Now let R consist of w and its permutants under $(1 \ 2 \ \ldots \ n)$. Then $G_n(w) = F/\bar{R}$, and $A_n(w)$ is trivial if and only if the composite mapping

$$\alpha: \bar{R} \overset{inc}{\to} F \overset{\pi}{\to} Z[t]/(t^n-1)$$

is onto, since both conditions are equivalent to $F = F'\bar{R}$. But Im α is just the subgroup generated by the set $R\alpha$, namely, by the cosets containing

$$f(t), tf(t), \ldots, t^{n-1}f(t) \ .$$

Thus, when $A_n(w)$ is trivial, we can find integers $b_0, \ldots, b_{n-1}$ such that

$$1 \equiv \sum_{i=0}^{n-1} b_i t^i f(t) \quad (\mathrm{mod}(t^n - 1)) \quad,$$

whereupon the polynomial $\sum_{i=0}^{n-1} b_i t^i$ is the desired inverse for $f(t)$, modulo $(t^n - 1)$.

Focussing our attention once more on the Fibonacci groups, we put $A(r,n) = F(r,n)^{ab}$, and recall that $r,n \in N$ with $r \geq 2$. Letting $f_r(t) = f(t)$ be the polynomial

$$f(t) = t^r - \sum_{i=0}^{r-1} t^i \quad,$$

we see from (4) that

$$|A(r,n)| = \pm \prod_{\xi^n = 1} f(\xi) \quad. \tag{7}$$

Theorem 5. $A(r,n)$ *is always a finite group.*

Proof.   By Theorem 4, we must show that the polynomial $f(t)$ of (7) can never vanish on a root of unity. Suppose for a contradiction that ξ lies on the unit circle and that $f(\xi) = 0$. Then

$$\xi^r = \sum_{i=0}^{r-1} \xi^i \quad,$$

and so, multiplying by $(1-\xi)$,

$$\xi^r - \xi^{r+1} = 1 - \xi^r \quad,$$

that is

$$2\xi^r = 1 + \xi^{r+1} \quad. \tag{8}$$

Taking moduli

$$1 + |\xi^{r+1}| = 2 = 2|\xi|^r = |1 + \xi^{r+1}| \quad,$$

which implies that ξ^{r+1} is real and non-negative. Hence $\xi^{r+1} = 1$ and by (8) $\xi^r = 1$ also. So $\xi = 1$ and thus

$$0 = f(\xi) = f(1) = 1 - r \quad,$$

which is the desired contradiction.

We proceed to compute the value of $|A(r,n)|$ exactly in certain special cases. To this end, let

$$r = kn + s \quad, \quad 0 \leq s < n \quad.$$

Theorem 6.

(i)   If $s \neq 1$, $|A(r,n)| = \dfrac{r-1}{s-1} |A(s,n)|$.

(ii)   If $s = 1$, $|A(r,n)| = n(r-1)$.

(iii) If $s = 0$, $|A(r,n)| = r - 1$.

(iv)   If $s = n-1$, $|A(r,n)| = (r-1) 2^{n-1}$.

(v)   If $s = n-2$, $|A(r,n)| = (r-1)(2^n+(-1)^{n+1})/3$.

Proof. All these formulae are consequences of equation (7), and their proofs depend on the fact that if $\xi^n = 1$, then

$$h(\xi) = \begin{cases} n \,, & \text{if } \xi = 1 \,, \\ 0 \,, & \text{otherwise} \,, \end{cases}$$

where $h(t)$ is the polynomial $1 + t + \ldots + t^{n-1}$. Thus,

$$|A(r,n)| = \pm (r-1) \prod_{\substack{\xi^n=1 \\ \xi \neq 1}} (1 + \xi + \ldots + \xi^{s-1} - \xi^s) \quad. \tag{9}$$

(i) follows at once from this. When $s = 1$, the right-hand side of (9) becomes

$$\pm (r-1) \prod_{\substack{\xi^n=1 \\ \xi \neq 1}} (1-\xi) = \pm (r-1)h(1) \quad,$$

which proves (ii). When $s = 0$, the sum $1 + \xi + \ldots + \xi^{s-1}$ in the right-hand side of (9) is empty, and (iii) follows. For (iv), note that

81

$$1 + \xi + \ldots + \xi^{n-2} - \xi^{n-1} = -2\xi^{n-1}$$

when $\xi^n = 1 \neq \xi$. As for (v),

$$1 + \xi + \ldots + \xi^{n-3} - \xi^{n-2} = -2\xi^{n-2} - \xi^{n-1} \quad ,$$

and the absolute value of this is just $|2 + \xi|$. The right-hand side of (9) is therefore equal to

$$\pm \ (r-1)h(-2) \quad ,$$

which yields assertion (v).

We now examine the behaviour of $c_n = |A(r,n)|$ for fixed r and increasing n .

Definition 4. If f, g are two monic polynomial in $Z[t]$, the complex number

$$g * f = \prod_{g(\xi)=0} f(\xi)$$

is called the *resolvent* of f and g .

It is clear that $g * f = \pm f * g$, and so formula (7) can be rewritten in the form

$$c_n = \prod_{f(\xi)=0} |\xi^n - 1| \quad ,$$

which we now use to prove our next result.

Theorem 7. *For* n *large enough,* $c_n < c_{n+1}$.

Proof. We examine the behaviour of the rational complex function

$$q_n(z) = \frac{z^{n+1} - 1}{z^n - 1}$$

as n tends to infinity. First of all, if $|z| > 1$,

$$|q_n(z)| \geq \frac{|z|^{n+1} - 1}{|z|^n + 1} = |z| - \frac{|z| + 1}{|z|^n + 1} \quad ,$$

which is increasing and exceeds 1 eventually. If on the other hand $|z| < 1$,

$$1 - |z|^n \leq |z^n - 1| \leq 1 + |z|^n \quad ,$$

so that $|q_n(z)|$ tends to 1 in this case. Now in the proof of Theorem 5, we showed that f has no root of modulus 1 , and so these are the only two cases that can arise in studying the limiting behaviour of

$$\frac{c_{n+1}}{c_n} = \prod_{f(\xi)=0} |q_n(\xi)| \quad .$$

Thus to prove the theorem, it is sufficient to show that f has at least one root outside the unit circle, and this follows at once from the fact that the product of its roots is ± 1 .

Theorem 8. *Given a group* G *, there are at most finitely many pairs* (r,n) *such that* $G \cong F(r,n)$.

Proof. If G/G' is infinite, the result follows from Theorem 5. So assume that G/G' is finite - of order m say - and suppose that $G = F(r,n)$. Adjunction of the relations $x_1 = x_2 = \ldots = x_n$ shows that G has the group $F(r,1)$ as a factor group and since $F(r,1) \simeq Z_{r-1}$, we must have $r \leq m+1$. So there are at most finitely many r such that $G \cong F(r,n)$, and for each of these, there are at most finitely many n with $c_n = m$, by Theorem 7. This proves the theorem.

EXERCISE 1.   Prove that $x_n \ldots x_1 = e$ in $F(2,n)$.

EXERCISE 2.   Prove that $[x_1,x_2]^2 = e$ in $F(2,n)$.

EXERCISE 3.   Prove that $F(n,n+1) \cong F(2n-1,n)$ for all $n \geq 2$.

EXERCISE 4. Show that $F(r,n) \cong Z_{r-1}$ when n is a divisor of r .

EXERCISE 5. Show that the cyclic group of order $2^{s-1} - 1$ appears at least s times among the $F(r,n)$.

EXERCISE 6. Show that for all $s \geq 1$, $F(2s+1,2)$ is a metacyclic group of order $4s(s+1)$.

EXERCISE 7. Prove that when $r \equiv 1 \pmod{n}$, $F(r,n)$ is a metacyclic group of order at most $n(r^n - 1)$.

EXERCISE 8. Prove that every finite group is a factor group of some $F(r,n)$.

EXERCISE 9. Define $F(r,n,k)$ to be the group $G_n(w)$ when w is the word $x_1 \ldots x_r x_{r+k}^{-1}$. Find necessary and sufficient conditions for $F(r,n,k)$ to have an infinite abelian factor group.

EXERCISE 10. Use the matrices

$$A_1 = \begin{pmatrix} -1 & 1 & 0 \\ -1 & 0 & 0 \\ 0 & 0 & 1 \end{pmatrix}, \quad A_2 = \begin{pmatrix} 0 & 1 & 0 \\ 0 & 0 & 1 \\ 1 & 0 & 0 \end{pmatrix}.$$

to show that the group $F(2,10)$ is infinite.

EXERCISE 11. Prove that the automorphism θ of $F(r,n)$ induced by $(12\ldots n)$ has order equal to n when n is sufficiently large relative to r .

EXERCISE 12. Prove that, in $F(2,n)$, we have the relation

$$x_1 \ldots x_n = e , \quad \text{or} \quad (x_1 \ldots x_n)^2 = e$$

according as n is even or odd.

4 · Presentations of subgroups

> Write the vision, and make it plain upon tables,
>
> that he may run that readeth it. (Habakkuk)

Suppose we are given a presentation $<X|R>$ for a group G . As might be expected, the derivation of a presentation for a specified factor group of G is easy enough (Theorem 3.3). By contrast, the corresponding problem for a subgroup H of G is no simple matter, and is in general very undecidable. As usual we dodge the pathology by confining ourselves to propitious cases, and describe the general method in §12. The whole thing hinges on the derivation of a certain Schreier transversal U , in terms of which we obtain free generators for H (as in Lemma 2.3). A simple trick gives relators for H in terms of the generators X of G and these must be rewritten as words in the free generators of H . If necessary, we can then perform Tietze transformations (Theorem 4.3) on the resulting presentation to reduce it to a more suitable form.

The method of calculating U varies according to what H is and how it is specified. For example, the method of §6 (see Example 6.2) contains an algorithm for computing U in the case where G is finitely generated and $H \supseteq G'$. The best general method, however, is that invented by J.A. Todd and H.S.M. Coxeter in 1936 and known as coset enumeration. It works in any specific situation when H is the subgroup generated by a set Y of words in X , provided only that $|X|$, $|R|$, $|Y|$ and $|G:H|$ are all finite. We describe it in §§10, 11.

The method of coset enumeration has had a highly successful career, particularly since the advent of high-speed computing machines, and has been used among other things for identifying new finite simple groups. The process has been implemented on many machines throughout the world, sometimes in conjunction with other

methods like those mentioned in the previous paragraph. In deference to this, we devote §13 to describing a refinement of the
method, due in this form to W.O.J. Moser, which yields a presentation for H by means of a single computation (by hand or
machine).

§10. A special case

The method described below can be applied to any *specific* fi-
nite presentation $G = <X|R>$, and always works provided $|G|$ is
finite. It not only yields such information as

the order of G ,

a faithful permutation representation of G ,

a Cayley diagram for G , and

a Schreier transversal for $\bar{R}$ in $F(X)$,

but is also great fun. This is how it goes.

For each relation $r = x_1 \ldots x_n \in R$, with $x_1 \ldots x_n$ a reduced
word in $X \cup X^{-1}$, we draw a rectangular table having $n + 1$
columns and a certain (for the moment unspecified) number of rows:

x_1	x_2	$\ldots$	x_n
1	2	$\ldots$	1
2		$\ldots$	2

We begin by entering the symbol 1 in the first and last places
of the first row of each table, the remaining places in the first
rows being as yet empty. We then pick an empty space next to
some 1 (either to the right or left of it) and fill it with the
symbol 2 . For the sake of definiteness, suppose the situation
to be as in the above diagram, with 2 immediately to the right
of 1 and $x_1 \in X \cup X^{-1}$ lying between them. We record the information $'1x_1 = 2'$ (and/or $'2x_1^{-1} = 1'$) in a monitor table, and
such an equation is eminently reasonable if we think of the numbers 1,2 as corresponding to the elements $e, x_1 \in G$, respectively. Now we put a 2 in the first and last places of the
second row of each table and, wherever in any table 1 lies to

86

the left of an empty space with x_1 between the two spaces, or
to the right of an empty space with x_1^{-1} between, we fill that
empty space with a 2 . Similarly, if 2 lies to the right (left)
of an empty space with $x_1(x_1^{-1})$ between, we fill that space with
a 1 . This purely mechanical process is known as *scanning*, and
it is here that most of the work is involved. Having made sure
that no more spaces can be filled in this way, we enter the symbol
3 in any empty space that is adjacent to a filled space. Having
recorded the corresponding information (of the form, 'ix = 3') in
our monitor table, we begin a new row in the relator tables and
scan as above. In similar fashion, we introduce the symbol 4 ,
record information, begin a fourth row, and scan again. We continue in this way until there are no more empty spaces, whereupon
the number of rows in each relator table is equal to $|G|$.

In order to reach the situation where all the tables are complete, we clearly need more information (of the form ix = j)
than is contained in our definitions of new symbols, and this is
supplied when any row of any table becomes complete. For suppose
we are in the position where such a row has but one remaining
empty space, and that space is filled as indicated in the following diagram:

$$\text{(1)}$$

Now this transition involves *two* pieces of information, namely

$$ix_\ell = j \ , \quad kx_{\ell+1}^{-1} = j \ ,$$

and we must distinguish between two cases. If this is the first
time the symbol j has appeared, one of these two equations is a
definition, and the other may be regarded as a bonus. On the
other hand, the arrival of j may be the result of scanning when
either one or both of these equations is already known. The lat-

ter case yields no new information, while in the former we obtain a bonus as before. There is one further possibility, but we postpone this in order to give a couple of examples.

Example 1. Take the cyclic group $G = \langle x \mid x^4 \rangle$ of order 4 . Here, there is only one relator table; it is headed $x\,x\,x\,x$ and has five columns:

x	x	x	x
1 ┤ 2 ┤ 3 ┤ 4 ⧧ 1
2	3	4 ⧧ 1	2
3	4 ⧧ 1	2	3
4 ⧧ 1	2	3	4

We have made the definitions

$$1x = 2 \, , \quad 2x = 3 \, , \quad 3x = 4 \, ,$$

and with the completion of the first row, have obtained as a bonus that $4x = 1$. For explanatory purposes, we have indicated this by dashes on the vertical lines of the table at the corresponding points - one dash for a definition, two for a bonus, and three when a row completes without yielding new information.

Since the table completes after 4 rows, we deduce that $|G| = 4$. Had we bothered to draw them, our monitor tables would have looked like this:

definition	bonus		x
1x = 2		1	2
2x = 3		2	3
3x = 4	4x = 1	3	4
		4	1

The Schreier transversal, which will play so vital a role in §12, comes from the 'definition' column of the first monitor table, and is $\{e,x,x^2,x^3\}$ in this case. The second monitor table yields the (regular) permutation representation $x \mapsto (1234) \in S_4$, and the corresponding Cayley diagram is as follows:

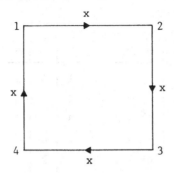

Example 2. As an example where the answer is not quite so obvious in advance, consider the Fibonacci group

$$G = F(2,3) = <a,b,c \mid ab = c , \quad bc = a , \quad ca = b> .$$

Eliminating c by Tietze transformations, we have

$$G = <a,b \mid baba^{-1} , \quad abab^{-1}> ,$$

and thus two relator tables. The reader can either check the working, or preferably do the enumeration himself using only the 'definition' column in the first monitor table below. If the same definitions are made in the same order, the result should look like this.

b	a	b	a^{-1}	
1	2	3	4	1
2	6	8	3	2
3	4	6	7	3
4	5	2	6	4
5	8	1	2	5
6	7	5	8	6
7	1	4	5	7
8	3	7	1	8

a	b	a	b^{-1}	
1	4	5	2	1
2	3	4	6	2
3	7	1	4	3
4	6	7	5	4
5	2	6	8	5
6	8	3	7	6
7	5	8	1	7
8	1	2	3	8

It follows that $F(2,3)$ has order 8. We have omitted the triple dashes here; as with the double dashes, the points where they appear are not always unique. The monitor tables are as follows:

definitions	bonus
1b = 2	
2a = 3	
3b = 4	1a = 4
4b = 5	5a = 2
2b = 6	4a = 6
3a = 7	6b = 7
	7b = 1
	7a = 5
6a = 8	8b = 3
	5b = 8
	8a = 1

	a	b
1	4	2
2	3	6
3	7	4
4	6	5
5	2	8
6	8	7
7	5	1
8	1	3

Returning to the general case, we continue the analysis of what occurs when a row completes, where the following alarming contingency remains to be considered. It may (and sometimes does) happen that in the process of scanning, the bonus information obtained from the completion of a row is inconsistent with what we already know. In terms of the diagram (1), this occurs when our monitor tables at this point contain the information

$$ix_\ell = j \ , \quad kx_{\ell+1}^{-1} = m \qquad \text{with } j \neq m \ ,$$

$$\text{or} \quad ix_\ell = j \ , \quad jx_{\ell+1} = n \qquad \text{with } k \neq n \ ,$$

$$\text{or} \quad mx_\ell^{-1} = h \ , \quad kx_{\ell+1}^{-1} = m \qquad \text{with } h \neq i \ .$$

In every case we obtain inconsistent information of the form $ix = j$, $ix = m$ (say), with $j \neq m$. This induces the phenomenon known as *coset collapse*, and we must proceed as follows. We conclude that $j = m$, and replace the larger of these numbers by the smaller throughout all our tables, noting that this converts the definition of the larger symbol into a bonus. This may yield extra bonuses treated in the usual way or further collapses, whereupon we pursue the same strategy, and continue in this way until all inconsistent information has been removed. We then delete all rows in the relator tables corresponding to the offending symbols (that is, the larger of each inconsistent pair). If so desired, we can now rename our symbols to form a set of consecutive natural numbers beginning with 1 . The interrupted scanning process can now continue, and we are back in the old routine. The situation is well illustrated by the following example.

Example 3.   We apply the method to the group

$$T_2 = \langle x,y \mid x^2y^3 , x^3y^4 \rangle ,$$

and continue until a collapse occurs (at ‡), whereupon the partially completed tables are as follows:

x	x	y	y	y	
1	2	3	4	5	1
2	3	6	3	4	2
3					3
4			3	4	
5			3	4	5
6					6

x	x	x	y	y	y	y	
1	2	3	6	3	4	5	1
2	3						2
3							3
4					3	4	
5				3	4	5	
6							6

	definition	bonus
	1x = 2	
	2x = 3	
	3y = 4	
* ...	4y = 5	5y = 1
	3x = 6	6y = 3
		4y = 2 ... *

	x	y
1	2	
2	3	
3	6	4
4		5
5		1
6		3

The inconsistent information (labelled *) tells us that
5 = 2 . We replace 5 by 2 and 6 by 5 throughout and continue scanning. As there are no further collapses, our relator
tables now appear as follows:

	x	x	y	y	y
1	2	3	4	2	1
2	3	5	3	4	2
3	5			5	3
4			5	3	4
5					5

	x	x	x	y	y	y	y
1	2	3	5	3	4	2	1
2	3	5		5	3	4	2
3	5					5	3
4					5	3	4
5							5

Defining $5x = 6$, the second row of the second table gives
$6y = 5$, and completion of the third row of the first table gives
the collapse $6 = 5$. Hence $5y = 5$ and together with the
equations

$$5y = 3 , \quad 3y = 4 , \quad 4y = 2 , \quad 2y = 1 ,$$

this leads to successive collapses $5 = 3 = 4 = 2 = 1$. *Since the
first row of both tables is complete*, we are not surprised to deduce that T is just the trivial group.

This example highlights the value of the strategy of defining
new symbols in such a way that the first rows of all the relator
tables become complete as quickly as possible. The retrospective
reader will recall that the same ploy was adopted in both of the
earlier examples. Another point worthy of note is provided by the
following generalization.

Example 4. It is plain that the group

$$T_n = \langle x,y \mid x^n y^{n+1} , x^{n+1} y^{n+2} \rangle , \quad n \in N ,$$

is trivial, and that the first definition must be of the form
$1x = 2$, or $1y^{-1} = 2$. In fact the definitions must continue to
be of the form $ix = j$ or $iy^{-1} = j$ (with $i < j$) for some time.

It is not hard to see that at least the first n definitions must
be of this type (see Exercise 2), and that no row can become complete before the nth definition is made. Thus, for any $n \in N$,
there is a presentation of the trivial group for which the coset
enumeration requires the use of more than n symbols. While the
process will always terminate for finite groups, there is in
general no way of bounding in advance the number of symbols that
will be needed. For this reason, the method cannot be described
as an algorithm.

   We conclude this section with an explanation of why the method
works, that is, we prove that if a set of tables is both complete
and consistent, then the resulting permutation representation is
the regular one. Thus, given a finite presentation

$$G = \langle x_1, \ldots, x_n \mid R \rangle \quad ,$$

let F be the free group on $X = \{x_1, \ldots, x_n\}$ and $\bar{R}$ the normal
closure of R in F . We assume that each $r \in R$ is a reduced
word in $X^{\pm 1}$, and that *every* x_j *appears in some such* r (see
Exercise 4). Now suppose that a coset enumeration concludes
consistently after g rows. This means that each column of each
relator table contains the symbols $1, \ldots, g$ in some order, and
so the above assumption implies that our second monitor table is
complete and contains the information

$$ix_k = j \ , \quad 1 \le i \le g, \ 1 \le k \le n \ , \tag{2}$$

for various values of j . Now by consistency,

$$i \ne i' \implies ix_k \ne i'x_k \ \text{for any} \ k \ ,$$

and we have a mapping θ from X to the symmetric group S_g
on $\{1, \ldots, g\}$. Since every row of every relator table begins and
ends with the same symbol, the extension θ' of θ to F maps
each $r \in R$ to e , and so (cf. the Substitution Test) $R \subseteq \text{Ker } \theta'$
and θ' induces a homomorphism

$$\rho: G \to S_g \quad .$$

We now examine more closely the information in our first monitor table. Its first column consists of $g - 1$ definitions, the jth of which has the form

$$ix = j + 1 \ , \quad 1 \le j \le g - 1 \ ,$$

with $i \le j$ and $x \in X \cup X^{-1}$. We use these equations to define inductively a set of words $u_1, \ldots, u_g \in F$ as follows. We put $u_1 = e$ and, for $1 < j \le g$, $u_{j+1} = u_i x$. A simple induction shows that

$$l(u_j \theta') = j \ , \quad 1 \le j \le g \ , \tag{3}$$

and our original assumption of consistency ensures that $u_1, \ldots, u_g$ have the Schreier property (see Exercise 5). Regarding the u_j as members of G , the equations (3) show that their images under ρ map 1 to each of $1, \ldots, g$ respectively, and it follows that ρ is transitive.

We must now show that ρ is faithful, that is, Ker $\theta' \subseteq \bar{R}$. This requires a more delicate analysis and, to simplify notation, we suppress the θ' and denote the corresponding action of F on $\{1, \ldots, g\}$ by juxtaposition. Recall that our first monitor table consists of the ng equations (2) (where the information $ix_k = j$ may appear in the equivalent form $jx_k^{-1} = i$), and note that we have exactly $(n-1)g + 1$ bonus equations, a familiar and significant number! We now claim that if the equation $ix = j$ appears in this table, then $u_i x = r u_j$ in F , for some $r \in \bar{R}$. This holds automatically for the definitions (with $r = e$), and for the rest, we must examine more closely what happens when a row completes.

We assume (inductively) that our claim is valid up to a given point in the enumeration, when a row completes as in the diagram (1). Suppose we are in the kth row of the relator table for

$r \in R$, so that the local picture of the completed row is as follows:

$$(4)$$

where the *bonus* information is $ix = j$, and r is the reduced word sxt . We examine in turn each of the three possibilities, where the equation $ix = j$ is either

(i)   known already, or

(ii)   not yet known but consistent, or

(iii) inconsistent.

In case (i), no entry is made in the monitor table and there is nothing to prove. In case (ii), we must justify the new entry $ix = j$ in the bonus column. Now by our inductive assumption, we know that

$$u_k s = r_1 u_i \ , \quad u_k t^{-1} = r_2 u_j \ , \tag{5}$$

where $r_1, r_2 \in \bar{R}$. It follows that:

$$
\begin{aligned}
u_i x &= r_1^{-1} u_k sx \\
&= r_1^{-1} (r_2 u_j t) sx \\
&= r_1^{-1} r_2 (u_j trt^{-1} u_j^{-1}) u_j \\
&= r_3 u_j \ ,
\end{aligned}
$$

say, with $r_3 \in \bar{R}$ as required. Passing to case (iii), we arrive at the same equation, but must pay heed to the fact that we already have an equation of the form

$$
\left.
\begin{aligned}
u_i x &= r_4 u_\ell \ , \quad j \neq \ell, \ r_4 \in \bar{R} \ , \\[1em]
\text{or} \quad u_h x &= r_5 u_j \ , \quad h \neq i, \ r_5 \in \bar{R} \ .
\end{aligned}
\right\}
\tag{6}
$$

In the former case, u_j and u_ℓ belong to the same coset of $\bar{R}$ in F, and so replacing ℓ by j (say) throughout will not violate our inductive assumption. For example, if

$$cy = \ell \implies u_c y = r_6 u_\ell \; ,$$

then

$$cy = j \implies u_c y = r_6 r_4^{-1} r_3 u_j \; .$$

The second possibility $(r_5^{-1} u_h = r_3^{-1} u_i)$ is equally harmless, as also are the subsequent collapses that may result.

We have thus proved the claim made at the end of the penultimate paragraph, and we proceed to deduce that $\mathrm{Ker}\,\theta' \subseteq \bar{R}$. If w is a reduced word in F with $w \in \mathrm{Ker}\,\theta'$, we have $1w = 1$, where w acts letterwise from left to right. In accordance with our claim, we deduce that

$$w = u_1 w = r u_1 = r \; , \tag{7}$$

for $r \in \bar{R}$ as required. This argument actually yields rather more, namely, if $w \in G$ and $w\rho$ fixes 1, then $w = e$. Since ρ is transitive, this implies that ρ is regular and $|G| = g$. Thus, provided the enumeration completes, all the claims made at the outset of this section have been proved.

EXERCISE 1.   Use the permutation representation given by the enumeration in Example 2 to identify $F(2,3)$. Can the corresponding Cayley diagram be embedded in the plane (or 2-sphere, or torus)?

EXERCISE 2.   Prove in detail that in enumerating cosets for the presentation

$$T_n = \langle x,y \mid x^n y^{n+1}, \; x^{n+1} y^{n+2} \rangle \; , \quad n \in \mathbb{N} \; ,$$

at least $n + 1$ symbols are needed. Can you enlarge this lower bound?

EXERCISE 3. Let $G = \langle X|R \rangle$ be a finitely presented group, and n a natural number. Find a presentation for G for which the coset enumeration requires more than n symbols.

EXERCISE 4. Let $G_1 = \langle X|R \rangle$ and, if Y is the set of those $x \in X$ such that $x^{\pm 1}$ appears as a letter in a member of R, let $G_2 = \langle Y|R \rangle$ and $F = F(X \backslash Y)$. Prove that $G_1 = G_2 * F$ and deduce that G_1 is infinite when $Y \neq X$.

EXERCISE 5. Prove that the words $u_1, \ldots, u_g$ defined on p. 94 satisfy (3), and that they have the Schreier property.

EXERCISE 6. Let F be the free group on $\{a, b\}$ and $R = \{baba^{-1}, abab^{-1}\} \subseteq F$. Use the first monitor table in Example 2 to write down a Schreier transversal for $\bar{R}$ in F (definition column) and compute a set of free generators for $\bar{R}$ (bonus column).

EXERCISE 7 (cf. Exercise 6.6(iii) and Definition 9.2). Use the method of coset enumeration to identify the Fibonacci group $F(2,5)$.

EXERCISE 8. Identify the von Dyck group $D(3,3,2)$.

EXERCISE 9. Prove that a coset enumeration on the free product

$$Z_2 * Z_2 = \langle x, y \mid x^2, y^2 \rangle$$

can never complete.

EXERCISE 10. Let $G = \langle X|R \rangle$ and $\bar{R}$ be the normal closure of R in $\langle X| \, \rangle$. Suppose a coset enumeration is consistent and complete after g rows and that $u_1, \ldots, u_g$ are as defined on p.94   Prove that $\bar{R}$ is freely generated by the set

$$\{u_i x u_j^{-1} \mid \text{'}ix = j\text{'} \text{ appears as a bonus}\} \quad .$$

§11. Coset enumeration

By means of a very simple adjustment, the method described in the preceding section can be carried out relative to a subgroup H of a finitely presented group $G = \langle X | R \rangle$. H is specified as the subgroup of G generated by a finite set Y of words in $X^{\pm 1}$, and the process terminates after a finite number of steps provided that $|G:H|$ is finite. We obtain such information as

the index of H in G ,

the permutation representation of G on the (right) cosets of H ,

the coset diagram of G relative to H , and

a transversal for H in G with the Schreier property.

While it yields less information than the old method in general, the advantage of the new method is that the tables become complete much sooner. When $H = E$, the two methods yield the same information.

At the outset we draw up relator and monitor tables as before, and in addition a table for each generator y of H . These new tables are constructed in the same way as the relator tables, with the letters of y separating adjacent columns, except that they have only one row, beginning and ending with the symbol 1 . The method then proceeds exactly as above, with the Y-tables being completed according to the same rules as the R-tables. The process again terminates when there are no more empty spaces, whereupon $|G:H|$ is just the number of rows in each R-table.

Example 1.   We begin with a simple example, to which the answer is already known. Letting $G = \langle x | x^6 \rangle$ and $H = \langle x^3 \rangle \leq G$, the tables are as follows:

	x	x	x	x	x	x	
1	2	3	1	2	3	≡	1
2	3	1	2	3	≡	1	2
3	1	2	3	≡	1	2	3

	x	x	x		
1	⊢	2	⊢	3	≡ 1

definition	bonus		x
1x = 2		1	2
2x = 3	3x = 1	2	3
		3	1

The one-rowed table is the first to complete, and the resulting
bonus is sufficient for the three rows of the relator table to com-
plete without yielding new information. We deduce that $|G:H| = 3$
and obtain the permutation representation $x \mapsto (123) \in S_3$ for G .
The Schreier transversal for H in G is just $\{e,x,x^2\}$, and
the coset diagram is as follows:

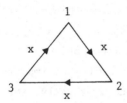

Example 2. We carry out the same process for the von Dyck group
$D(3,3,2)$ (see Exercise 10.8):

$$G = \langle x,y \mid x^3, y^3, (xy)^2 \rangle \ ,$$

with respect to the subgroup $H = \langle x \rangle$. The tables are as follows:

	x	x	x	
1	1	1	≡	1
2	3	4	4	2
3	4	2		3
4	2	3		4

	y	y	y	
1	2	3	≡	1
2	3	1		2
3	1	2		3
4	4	4	≡	4

	x	y	x	y	
1	1	2	3		1
2	3	1	1		2
3	4	4	2		3
4	2	3	4		4

	x	
1	≡	1

definition	bonus		x	y
	$1x = 1$	1	1	2
$1y = 2$		2	3	3
$2y = 3$	$3y = 1$	3	4	1
	$2x = 3$	4	2	4
$3x = 4$	$4x = 2$			
	$4y = 4$			

We deduce that $|G : H| = 4$, and that $\{e,y,y^2,y^2x\}$ is a Schreier transversal. We also have the permutation representation

$$\left.\begin{array}{l} \alpha: G \to S_4 \\[4pt] \quad x \mapsto (234) \\[4pt] \quad y \mapsto (123) \end{array}\right\}$$

and we deduce that $|x| \geq |x\alpha| = 3$. Since x^3 is a relator, we already know that $|x| \leq 3$, and conclude that $|H| = |x| = 3$. It follows that $|G| = 12$, and since $A_4 = \langle(234),(123)\rangle = \text{Im } \alpha$, we have identified G up to isomorphism.

This example illustrates a number of important points of which some are typical and others not. Firstly, the situation where the generators of H form a subset of the generators of G (that is, $Y \subseteq X$) is a particularly propitious one. In such cases, we often omit the one-rowed tables altogether and simply subsume the corresponding information into the second column of the first monitor table. Next, the periodic nature of the relators results in some redundancies in the relator tables. When this happens, it saves time to replace the internal entries of a superfluous row by a dash linking the first and last entries as in Example 3 below. Finally, the fact that G has been identified, that is, α is faithful, is by no means typical (cf. Example 1 and Exercise 9) and must be regarded simply as a stroke of good luck. The next example illustrates each of these points, and we counsel the reader to carry out the computation for himself.

Example 3.   We enumerate cosets with respect to H = <x> in
D(4,3,2) :

$$G = <x,y \mid x^4, y^3, (xy)^2> \quad .$$

x	x	x ·	x	
1	1	1	1	1
2	3	4	5	2
3				3
4				4
5				5
6	6	6	6	6

y	y	y	
1	2	3	1
2			2
3			3
4	5	6	4
5			5
6			6

x	y	x	y	
1	1	2	3	1
2				2
3	4	5	2	3
4	5	6	6	4
5				5
6				6

definition	bonus		x	y
	1x = 1	1	1	2
1y = 2		2	3	3
2y = 3	3y = 1 2x = 3	3	4	1
3x = 4		4	5	5
4x = 5	5x = 2 4y = 5	5	2	6
5y = 6	6y = 4 6x = 6	6	6	4

By the same argument as used in the previous example, we deduce
that |G| = 24 , and it is not hard to show (Exercise 3) that
$G \cong S_4$. We leave it to the reader to work out the other by-
products of the computation.

Example 4.   To illustrate the phenomenon of coset collapse in the
adapted method, we consider the group

$$F(2,5) = <x,a,b,c,d \mid xa = b, \; ab = c, \; bc = d, \; cd = x, \; dx = a> \; ,$$

and enumerate cosets with respect to <x> .

x	a	b⁻¹	
1	1	3	1
2	3		2
3		2	3

a	b	c⁻¹	
1	3	2	1
2	1	3	2
3			3

b	c	d⁻¹	
1	3	2	1
2		1	2
3	2	3	3

c	d	x⁻¹	
1	2	1	1
2	3	3	2
3			3

d	x	a⁻¹	
1	2	3	1
2	1	1	2
3	3		3

definition	bonus			
	$1x = 1$			
$1c = 2$	$2d = 1$	$2a = 1$		
$1a = 3$	$1b = 3$	$3b = 2$	$2c = 3$	$3d = 3$
	$2x = 3$	$1d = 2$	$3c = 2$	

	x	a	b	c	d
1	1	3	3	2	2
2	3	1		3	1
3			2		3

At this point we deduce that $1 = 2c^{-1} = 3$, whence
$3 = 1b = 3b = 2$, and the tables collapse to one row each. The
second monitor table tells us that each of the five generators
fixes 1 , so that $F(2,5) = \langle x \rangle$ and is thus abelian. Since we
already know (Exercise 6.6(iii)) that the derived factor group of
$F(2,5)$ is Z_{11} , we deduce that $F(2,5) \cong Z_{11}$ (as in the table
of §9).

Example 5. As our final example, we take another cyclically
presented group

$$G = \langle a,b,c \mid abcab^{-1}, bcabc^{-1}, cabca^{-1} \rangle$$

and enumerate cosets with respect to the subgroup $H = \langle abc \rangle$.

a	b	c	a	b⁻¹	
1	2	3	1	2	1
2	3	1	2	3	2
3	1	2	3	1	3

b	c	a	b	c⁻¹	
1	2	3	1	2	1
2	3	1	2	3	2
3	1	2	3	1	3

c	a	b	c	a⁻¹	
1	2	3	1	2	1
2	3	1	2	3	2
3	1	2	3	1	3

a	b	c	
1	2	3	1
2	3	1	2
3	1	2	3

definition	bonus		
$1a = 2$			
$2b = 3$	$3c = 1$	$1b = 2$	$2c = 3$
	$3a = 1$	$3b = 1$	$2a = 3$
	$1c = 2$		

	a	b	c	a⁻¹	b⁻¹	c⁻¹
1	2	2	2	3	3	3
2	3	3	3	1	1	1
3	1	1	1	2	2	2

Thus the tables complete consistently after only two definitions and we deduce that $|G:H| = 3$. The three rows involving the symbol ‡ will play a significant role later, as will the self-explanatory extension of the second monitor table.

As in the previous section, we conclude with a proof that when the process completes, it yields correct information. Thus we let $G = <x_1,\ldots,x_n \mid R>$ and assume as above that each x_i appears in some member of R or Y . We let Y be a finite set of generators for H and assume that the process is consistent and complete, with g rows in each R-table. Just as in the previous section, we obtain a transitive permutation representation $\rho: G \to S_g$, and a Schreier set $u_1,\ldots,u_g$. Under the resulting action of G on $\{1,\ldots,g\}$, we see that each $y \in Y$ fixes the symbol 1 ; this is just the content of the one-rowed Y-tables. Letting $K = \{k \in G \mid 1k = 1\} \le G$ be the stabilizer of 1 , we see that $H \subseteq K$. To prove the reverse inclusion, we prove the analogue of the crucial claim on p.94.

Thus we claim that if the equation $ix = j$ appears in the first monitor table, then $u_i x = rhu_j$ for some $r \in \bar{R}$ and some word h in $Y \subseteq <X| >$. As before, if the equation $ix = j$ is a definition, then $w_i x = w_j$ and there is no problem. We again assume that the claim is valid up to a given point when a row completes, and assume in the first place that the information $ix = j$ accrues from the completion of the kth row of the R-table headed $r = sxt$, as in 10.(4). When this information is consistent, we argue as before, using the equations

$$u_k s = r_1 h_1 u_i \quad , \quad u_k t^{-1} = r_2 h_2 u_j$$

which hold by induction and correspond to 10.(5). We compute that

$$u_i x = h_1^{-1} r_1^{-1} u_k sx$$

$$= h_1^{-1} r_1^{-1} r_2 h_2 u_j tsx$$

$$= h_1^{-1}(r_1^{-1}r_2)h_1 \cdot (h_1^{-1}h_2u_j t)\, sxt(t^{-1}u_j^{-1}h_2^{-1}h_1) \cdot h_1^{-1}h_2 \cdot u_j \quad ,$$

$$= r_3 h_3 u_j \quad ,$$

which is of the required form, since $sxt = r \in R$. When this
information is inconsistent, we already have an equation of the
form

$$u_i x = r_4 h_4 u_\ell \ , \quad j \neq \ell \ ,$$

say, whereupon u_j and u_ℓ belong to the same right coset of the
subgroup $\bar{R}{<}Y{>}$ of $\langle X | \ \rangle$. It is thus clear that our inductive
assumption is not violated by the collapsing procedure. The
situation corresponding to the second alternative of 10.(6) is
treated in a similar way, as are additional collapses. We are
left with the situation where the bonus comes from the completion
of a one-rowed table. If this is headed $y = sxt \in Y$, then we
have

$$u_1 s = r_1 h_1 u_i \ , \quad u_1 t^{-1} = r_2 h_2 u_j \quad .$$

Since $u_1 = e$ (by definition), we have

$$u_i x = h_1^{-1} r_1^{-1} sx$$

$$= h_1^{-1} r_1^{-1} yt^{-1}$$

$$= h_1^{-1} r_1^{-1} yr_2 h_2 u_j$$

$$= (h_1^{-1}r_1^{-1}h_1)(h_1^{-1}yr_2y^{-1}h_1)(h_1^{-1}yh_2)u_j$$

$$= r_3 h_3 u_j \quad ,$$

as required, and we repeat the above argument.

Finally, assume that some word $w \in G$ fixes the symbol 1 .
Because of the claim just proved, we have

$$w = u_1 w = rhu_1 = rh \in \bar{R}<Y> \quad ,$$

regarding w as an element of F (cf. 10.(7)). This just means that, as an element of G , $w \in <Y> = H$. It follows that $K \subseteq H$ and that ρ is just the permutation prepresentation of G on the right cosets of H . Hence, $u_1, \ldots, u_g$ comprise a transversal for H in G , and $|G:H| = g$ as required.

EXERCISE 1.   Draw a coset diagram for the enumeration carried out in Example 2.

EXERCISE 2.   Describe the Schreier transversal and permutation representation resulting from Example 3, and draw the coset diagram.

EXERCISE 3.   In S_4 , find a 4-cycle and a 3-cycle which generate the group and whose product has order 2 . Deduce from the result of Example 3 that $D(4,3,2) \cong S_4$.

EXERCISE 4.   Prove that $D(5,3,2) \cong A_5$.

EXERCISE 5.   Enumerate the cosets of the subgroup $H = <xy,zxyz>$ in the group

$$G = <x,y,z \mid x^2,y^2,z^2,(xy)^3,(yz)^3,[x,z]> \quad .$$

EXERCISE 6.   Identify the group

$$<x,a,b,c,d \mid x = bd, \ a = cx, \ b = da, \ c = xb, \ d = ac> \quad .$$

EXERCISE 7.   Compute the order of the group

$$<x,y \mid x^4 = y^4 = e, \ yx = x^2y^2> \quad .$$

EXERCISE 8.   Find the index of $<x>$ in the group

$$G = \langle x, y \mid x^2yxy^3, y^2xyx^3 \rangle \quad .$$

EXERCISE 9.   Let ρ be the permutation representation of a group G obtained by enumerating cosets with respect to a subgroup H . Prove that Ker ρ is equal to the intersection of the conjugates of H . (This is the largest normal subgroup of G lying in H , and is called the *core* of H .)

EXERCISE 10.   Given a complete enumeration of the cosets of $H = \langle Y \rangle$ in G , prove that H is normal in G if and only if every $y \in Y$ fixes every symbol.

EXERCISE 11 (cf. Exercise 10.10).   Prove that the $(n-1)g + 1$ words ry accruing from bonuses $u_i x = r y u_j$ $(x \in X)$ comprise a set of free generators of $\bar{R}\langle Y \rangle$.

§12. The Reidemeister-Schreier rewriting process

The main ingredients involved in writing down a presentation of a subgroup are already contained in the proof of the Nielsen-Schreier theorem carried out in §2.   To describe the general method, we fix some notation.   Thus, we let

$$F = F(X) \; , \quad G = \langle X \mid R \rangle = F/\bar{R} \; ,$$

and, for a given subgroup H of G , we let K be the preimage of H under the natural map: $F \to G$, so that $H = K/\bar{R}$.   The subgroup H may be specified (as in the previous section) as the subgroup of G generated by a given set of words $Y = Y(X)$, or in some other way, for example, as some canonical subgroup of G such as G' .   We further suppose that a Schreier transversal U for K in F is already to hand.   We have so far discussed two ways of computing U ; the method of the previous section always works provided X, R, Y and $|G : H|$ are all finite, while the methods of §6 apply whenever X is finite and $H \supseteq G'$.

As in §2, we let $\bar{f}$ denote the member of U representing the coset Kf (f $\in$ F) , so that the set

$$B = \{ux \, \overline{ux}^{-1} \mid u \in U, \, x \in X, \, ux \notin U\}$$

forms a set of free generators for K . Now each member of the
set

$$\hat{R} = \{uru^{-1} \mid u \in U, \, r \in R\}$$

belongs to $\overline{R} \subseteq K$, and can thus be written (uniquely) as a word
in the elements of B (as in Lemma 2.3). Denoting the resulting
set by $\hat{S}$, the transition from $\hat{R}$ to $\hat{S}$ is known as the
Reidemeister-Schreier rewriting process.

Theorem 1.   *With the above notation,* $<B|\hat{S}>$ *is a presentation*
for H .

Proof.   In view of the Hasse diagram

we merely have to prove that $\overline{R}$ is the normal closure of $\hat{S}$
in K . Since $\hat{S} \subseteq \overline{R} \lhd K$, one inclusion is obvious, while for
the other, we first note that any f in F can be written as
f = ku with $k \in K$, u = $\overline{f} \in U$. Hence, a typical generator
frf^{-1} of $\overline{R}$ is equal to $k(uru^{-1})k^{-1}$, which belongs to the
normal closure in K of $\hat{R} = \hat{S}$ as required.

The presentation for H given by this theorem is usually very
unwieldy, and we often make use of Tietze transformations to re-
duce it either to a more amenable form, or to the form $<Y|S>$

(in the case where H is originally specified as the subgroup of G generated by Y). We illustrate both cases directly, and then go on to give a number of further examples of general interest.

Example 1.   As usual, we begin with a group where the answer is known in advance, and find a presentation for the subgroup $H = \langle x \rangle$ in the group

$$G = \langle x,y \mid x^3, y^3, (xy)^2 \rangle \ .$$

Referring to Example 11.2, we see that $U = \{e, y, y^2, y^2x\}$, and use the bonus column of the first monitor table to compute the free generators B of K . These appear in the left-hand part of the following table, while the right-hand part simply lists the members of $\hat{R}$.

U \ X	x	y			
e	x	$-$	x^3	y^3	$(xy)^2$
y	yxy^{-2}	$-$	yx^3y^{-1}	y^3	$(yx)^2$
y^2	$-$	y^3	$y^2x^3y^{-2}$	y^3	y^2xyxy^{-1}
y^2x	$y^2x^2y^{-1}$	$y^2xyx^{-1}y^{-2}$	$y^2x^3y^{-2}$	$y^2xyx^{-1}y^{-2}$	$y^2x^2yxyx^{-1}y^{-2}$

Thus, for example, the last entry in the left-hand part of the table comes from the last entry in the bonus column on p.100, which asserts that $(y^2x)y = y^2x$. We now redraw this table, renaming the generators (in chronological order), and rewriting the relations in terms of them as described above. While Lemma 2.3 gives the algorithm for this process, it is often simpler and more satisfying to do it by inspection.

B		Ŝ		
b_1	$-$	b_1^3	b_2	$b_1b_3b_2$
b_3	$-$	b_3b_4	$-$	$b_3b_2b_1$
$-$	b_2	b_4b_3	$-$	b_5b_4
b_4	b_5	$-$	b_5^3	b_4b_5

Now our original generator x is just b_1, so we use Tietze transformations to eliminate the other generators, beginning with b_2 :

$$H = \langle b_1, b_3, b_4, b_5 \mid b_1^3, b_1 b_3, b_3 b_4, b_4 b_5, b_5^3 \rangle \quad .$$

Next we eliminate b_3, b_4, b_5 to obtain

$$H = \langle b_1 \mid b_1^3, b_1^{-3} \rangle = \langle x \mid x^3 \rangle \quad ,$$

as expected.

Example 2. Next we give an example where the answer is not so obvious, and find a presentation for G', where

$$G = \langle x, y \mid x^2 yxy^3, y^2 xyx^3 \rangle \quad .$$

We see that

$$G/G' = \langle x, y \mid y^{-1} = x^3 y^3 = x^{-1} \rangle = \langle x \mid x^7 \rangle \quad ,$$

and we may take $U = \{x^i \mid 0 \le i \le 6\}$. Working modulo G', we easily compute the entries of our $B|\hat{R}$-table, and the result is as follows:

$U \backslash X$	x	y		
e	$-$	yx^{-1}	$x^2 yxy^3$	$y^2 xyx^3$
x	$-$	xyx^{-2}	$x^3 yxy^3 x^{-1}$	$xy^2 xyx^2$
x^2	$-$	$x^2 yx^{-3}$	$x^4 yxy^3 x^{-2}$	$x^2 y^2 xyx$
x^3	$-$	$x^3 yx^{-4}$	$x^5 yxy^3 x^{-3}$	$x^3 y^2 xy$
x^4	$-$	$x^4 yx^{-5}$	$x^6 yxy^3 x^{-4}$	$x^4 y^2 xyx^{-1}$
x^5	$-$	$x^5 yx^{-6}$	$x^7 yxy^3 x^{-5}$	$x^5 y^2 xyx^{-2}$
x^6	x^7	$x^6 y$	$x^8 yxy^3 x^{-6}$	$x^6 y^2 xyx^{-3}$

The following table describes the action of x on the b_i; the compilation of such a table frequently assists in the task of expressing the relators $\hat{R}$ in terms of the generators B.

b_i	xb_ix^{-1}
$b_1 = x^7$	b_1
$b_2 = yx^{-1}$	b_3
$b_3 = xyx^{-2}$	b_4
$b_4 = x^2yx^{-3}$	b_5
$b_5 = x^3yx^{-4}$	b_6
$b_6 = x^4yx^{-5}$	b_7
$b_7 = x^5yx^{-6}$	$b_8b_1^{-1}$
$b_8 = x^6y$	b_1b_2

We thus obtain that

$$s_1 = x^2yxy^3 = b_4b_6b_7b_8$$
$$s_2 = x^3yxy^3x^{-1} = b_5b_7b_8b_2$$
$$s_3 = x^4yxy^3x^{-2} = b_6b_8b_2b_3$$
$$s_4 = x^5yxy^3x^{-3} = b_7b_1b_2b_3b_4$$
$$s_5 = x^6yxy^3x^{-4} = b_8b_3b_4b_5$$
$$s_6 = x^7yxy^3x^{-5} = b_1b_2b_4b_5b_6$$
$$s_7 = x^8yxy^3x^{-6} = b_1b_3b_5b_6b_7$$
$$s_8 = y^2xyx^3 = b_2b_3b_5b_1$$
$$s_9 = xy^2xyx^2 = b_3b_4b_6b_1$$
$$s_{10} = x^2y^2xyx = b_4b_5b_7b_1$$
$$s_{11} = x^3y^2xy = b_5b_6b_8$$
$$s_{12} = x^4y^2xyx^{-1} = b_6b_7b_1b_2$$

$$s_{13} = x^5 y^2 xyx^{-2} = b_7 b_8 b_3$$
$$s_{14} = x^6 y^2 xyx^{-3} = b_8 b_2 b_4$$

We proceed to reduce this 8-generator 14-relator presentation to an easily recognizable form. Relators s_{11}, s_{13}, s_{14} respectively yield

$$b_5 = b_8^{-1} b_6^{-1}, \quad b_3 = b_8^{-1} b_7^{-1}, \quad b_4 = b_2^{-1} b_8^{-1},$$

which when substituted in s_6 , give

$$b_1 = b_8^2 .$$

Substitution of this in s_{12} yields

$$b_2 = b_8^{-2} b_7^{-1} b_6^{-1} , \quad \text{and so} \quad b_4 = b_6 b_7 b_8 .$$

We can now apply Tietze transformations to eliminate the superfluous generators b_1, b_2, b_3, b_4, b_5, and relators $s_6, s_{11}, s_{12}, s_{13}, s_{14}$. The remaining relators (in order) turn out to be:

$$s_1 = b_6 b_7 b_8 b_6 b_7 b_8$$
$$s_2 = b_8^{-1} b_6^{-1} b_7 b_8 b_8^{-2} b_7^{-1} b_6^{-1}$$
$$s_3 = b_6 b_8 b_8^{-2} b_7^{-1} b_6^{-1} b_8^{-1} b_7^{-1}$$
$$s_4 = b_7 b_8^2 b_8^{-2} b_7^{-1} b_6^{-1} b_8^{-1} b_7^{-1} b_6 b_7 b_8$$
$$s_5 = b_8 b_8^{-1} b_7^{-1} b_6 b_7 b_8 b_8^{-1} b_6^{-1}$$
$$s_7 = b_8^2 b_8^{-1} b_7^{-1} b_8^{-1} b_6^{-1} b_6 b_7$$
$$s_8 = b_8^{-2} b_7^{-1} b_6^{-1} b_8^{-1} b_7^{-1} b_8^{-1} b_6^{-1} b_8^2$$
$$s_9 = b_8^{-1} b_9^{-1} b_6 b_7 b_8 b_6 b_8^2$$
$$s_{10} = b_6 b_7 b_8 b_8^{-1} b_6^{-1} b_7 b_8^2$$

Relators s_5 and s_7 respectively assert that b_7 commutes with b_6 and b_8 , whereupon s_4 yields that b_6 and b_8 commute with each other, proving that G' is abelian. The remaining relators yield the following relation matrix for G' :

$$
\begin{pmatrix}
2 & 2 & 2 \\
-2 & 0 & -2 \\
0 & -2 & 0 \\
-2 & -2 & -2 \\
2 & 0 & 2 \\
0 & 2 & 2
\end{pmatrix}
$$

Row operations alone suffice to reduce this to the matrix

$$
\begin{pmatrix}
2 & 0 & 0 \\
0 & 2 & 0 \\
0 & 0 & 2
\end{pmatrix} \quad ,
$$

which proves that

$$
G' \cong Z_2 \times Z_2 \times Z_2 \quad ,
$$

whence $|G'| = 8$ and $|G| = 56$. For purposes of comparison we shall tackle this group by a different method in the next section.

Example 3.   As a somewhat simpler example, we find a presentation for G' when G is the von Dyck group

$$
D(3,3,3) = \langle x,y \mid x^3, y^3, (xy)^3 \rangle \quad .
$$

We see at once that $G^{ab} = \langle x \mid x^3 \rangle \times \langle y \mid y^3 \rangle$, so that $U = \{x^i y^j \mid 0 \le i,\ j \le 2\}$ will do for our Schreier transversal. Our $B|\hat{R}$ table is again compiled by working modulo G/G' , and the result is as follows:

112

	x	y			
e	–	–	x^3	y^3	$(xy)^3$
x	–	–	–	xy^3x^{-1}	$x^2yxyxyx^{-1}$
x^2	x^3	–	–	$x^2y^3x^{-2}$	$x^3yxyxyx^{-2}$
y	$yxy^{-1}x^{-1}$	–	yx^3y^{-1}	–	$(yx)^3$
xy	$xyxy^{-1}x^{-2}$	–	$xyx^3y^{-1}x^{-1}$	–	–
x^2y	x^2yxy^{-1}	–	$x^2yx^3y^{-1}x^{-2}$	–	–
y^2	$y^2xy^{-2}x^{-1}$	y^3	$y^2x^3y^{-2}$	–	$y^2xyxyxy^{-1}$
xy^2	$xy^2xy^{-2}x^{-2}$	xy^3x^{-1}	$xy^2x^3y^{-2}x^{-1}$	–	$xy^2(xy)^3y^{-2}x^{-1}$
x^2y^2	$x^2y^2xy^{-2}$	$x^2y^3x^{-2}$	$x^2y^2x^3y^{-2}x^{-2}$	–	$x^2y^2(xy)^3y^{-2}x^{-2}$

We rename the generators and rewrite the relators to obtain the following table:

B		$\hat{S}$		
–	–	0	7	267
–	–	–	8	348
0	–	–	9	0159
1	–	123	–	1590
2	–	231	–	–
3	–	312	–	–
4	7	456	–	483
5	8	564	–	5901
6	9	645	–	672

We have abbreviated the generators to their subscripts (and would, had it been necessary, have replaced i^{-1} by $\bar{i}$). This yields the presentation

$$G' = \langle 1,2,3,4,5,6 \mid 26,34,15,123,456 \rangle \quad .$$

Using the first four relators to eliminate 6,4,5,3, the last relator becomes $12\bar{1}\bar{2}$, and we deduce that $G' \cong Z \times Z$, the free abelian group of rank 2 .

Example 4. To show that the method also works for non-metabelian groups, we find a presentation for G' when G is the modular group

$$Z_2 * Z_3 = <x,y \mid x^2, y^3> \quad .$$

We take $U = \{x^i y^j \mid i = 0,1, \; j = 0,1,2\}$, so that our $B|\hat{R}$ table comes out as follows:

U	x	y		
e	$-$	$-$	x^2	y^3
y	$yxy^{-1}x^{-1}$	$-$	yx^2y^{-1}	$-$
y^2	$y^2xy^{-2}x^{-1}$	y^3	$y^2x^2y^{-2}$	$-$
x	x^2	$-$	$-$	xy^3x^{-1}
xy	$xyxy^{-1}$	$-$	$xyx^2y^{-1}x^{-1}$	$-$
xy^2	xy^2xy^{-2}	xy^3x^{-1}	$xy^2x^2y^{-2}x^{-1}$	$-$

Renaming and rewriting as in the previous example, the $B|\hat{S}$ table comes out like this:

B		Ŝ	
$-$	$-$	2	5
0	$-$	03	$-$
1	5	14	$-$
2	$-$	$-$	6
3	$-$	30	$-$
4	6	41	$-$

The generators 2,5,6 disappear first, closely followed by $3 = \bar{0}$ and $4 = \bar{1}$. We are left with two generators and no relators, so that G' is free of rank two in this case.

Definition 1. For $\ell, m, n \in Z$, we define the *triangle group* $\Delta(\ell, m, n)$ by the presentation:

$$\Delta(\ell,m,n) = \langle a,b,c \mid a^2,b^2,c^2,(ab)^\ell,(bc)^m,(ca)^n\rangle .$$

To avoid degeneracy and duplication, we usually assume that $\ell \geq m \geq n \geq 2$.

Example 5.   Consider the subgroup H of $\Delta(\ell,m,n)$ generated by $x = ab$ and $y = bc$. Since

$$x^a = ba = x^{-1}, \; x^b = b^{-1}ab^2 = x^{-1}, \; x^c = cabc = y^{-1}x^{-1}y,$$

$$y^a = abca = xy^{-1}x^{-1}, \; y^b = cb = y^{-1}, \; y^c = c^{-1}bc^2 = y^{-1},$$

we see that H is normal in G . Hence, by Theorem 3.3,

$$G/H = \langle a,b,c \mid a^2,b^2,c^2,(ab)^\ell,(bc)^m,(ca)^n,ab,bc\rangle$$

$$= \langle a,b,c \mid a^2,ab,ac\rangle \cong Z_2 ,$$

and we can take $\{e,a\}$ for our Schreier transversal U .   The $B\hat{|R}$- and $B\hat{|S}$- tables are then as follows:

U	a	b	c						
e	–	ba^{-1}	ca^{-1}	a^2	b^2	c^2	$(ab)^\ell$	$(bc)^m$	$(ca)^n$
a	a^2	ab	ac	–	ab^2a^{-1}	ac^2a^{-1}	$a(ab)^\ell a^{-1}$	$a(bc)^m a^{-1}$	$(ac)^n$

B			$\hat{S}$					
–	2	4	1	23	45	3^ℓ	$(25)^m$	$(41)^n$
1	3	5	–	32	54	$(12)^\ell$	$(34)^m$	5^n

Writing $x = ab = 3$, $z = ac = 5$, we have

$$H = \langle x,z \mid x^\ell,(x^{-1}z)^m,z^n\rangle .$$

In terms of our original generators x , $y=bc=x^{-1}z$, we see that H is just the von Dyck group $D(\ell,m,n)$.   It follows that

115

every triangle group contains the corresponding von Dyck group as a subgroup of index 2 .

Both the groups $\Delta(\ell,m,n)$ and $D(\ell,m,n)$, as well as the symmetric group S_n , are examples of *generalized Coxeter groups*, defined as follows.

Definition 2. Given a symmetric $n \times n$ matrix $M = (m_{ij})$ over the non-negative integers, consider the group

$$G(M) = \langle x_1,\ldots,x_n \mid R,S \rangle \quad ,$$

where

$$R = \{x_i^{m_{ii}} \mid 1 \le i \le n\}, \quad S = \{(x_i x_j)^{m_{ij}} \mid 1 \le i < j \le n\} \quad .$$

$G(M)$ is called the *generalized Coxeter group* determined by M . (N.B. Coxeter groups are given by the special case when each m_{ii} is 2 , as in S_n , for example.)

Example 6. It is clear that for $n \ge 2$, S_n is the Coxeter group determined by the $(n-1) \times (n-1)$ matrix M_n with super-diagonal and sub-diagonal entries equal to three, and with all other entries equal to 2 . We proceed to apply our method to compute a presentation for A_n . We complete the $B \mid \hat{R}$ -table using the transversal $U = \{e, x_1\}$, indicating the subscript ranges in the first row.

U	x_i	$x_i \ (2 \le i)$		$2 \le i$		$2 \le i$
e	$-$	$x_i x_1^{-1}$	x_1^2	x_i^2	$(x_1 x_2)^3$	$(x_i x_{i+1})^3$
x_1	x_1^2	$x_1 x_i$	$-$	$x_1 x_i^2 x_1^{-1}$	$x_1 (x_1 x_2)^3 x_1^{-1}$	$x_1 (x_i x_{i+1})^3 x_1^{-1}$

$3 \le j$	$2 \le i < j-1$
$(x_1 x_j)^2$	$(x_i x_j)^2$
$x_1 (x_1 x_j)^2 x_1^{-1}$	$x_1 (x_i x_j)^2 x_1^{-1}$

The B$|\hat{S}$ table is completed by renaming and rewriting in the usual way, with subscript ranges as above.

B				$\hat{S}$			
$-$	y_i	t	$y_i z_i$	z_2^3	$(y_i z_{i+1})^3$	z_i^2	$(y_i z_j)^2$
t	z_i	$-$	$z_i y_i$	$(ty_2)^3$	$(z_i y_{i+1})^3$	$(ty_j)^2$	$(z_i y_j)^2$

Eliminating t and replacing z_i by y_i^{-1} $(2 \leq i \leq n-1)$, we procure the presentation

$$A_n = \langle y_2, \ldots, y_{n-1} \mid y_2^3, \{(y_i y_{i+1}^{-1})^3 \mid 2 \leq i < n-1\} \, ,$$

$$\{y_j^2 \mid 3 \leq j \leq n-1\}, \{(y_i y_j^{-1})^2 \mid 2 \leq i < j-1 < n-1\}\rangle \, .$$

Because of the second set of relators, the $"^{-1}\text{'s}"$ can be deleted from the first and third sets. We deduce that A_n is just the generalized Coxeter group determined by the matrix obtained from M_{n-1} by adding 1 to its first entry.

We conclude this section with the observation that under suitable circumstances, the process described above can be used to compute the multiplicator $M(G)$ of a finite group G (see the introduction to Chapter III). Thus, if $G = \langle X|R \rangle$ and $\bar{R}$ is the normal closure of R in $F = F(X)$, we have

$$\frac{F}{[F,\bar{R}]} = \langle X|[X,R] \rangle \, .$$

Given a Schreier transversal for $\bar{R}$ in F , we find a presentation for the subgroup $\langle R \rangle = \bar{R}/[F,\bar{R}]$ of this group. This subgroup is abelian, and $M(G)$ is just its torsion subgroup, which can now be identified by the methods of §6. It turns out however that even for relatively innocent-looking groups (see Exercise 9), the calculation is somewhat horrendous.

EXERCISE 1.   Let G be the group

$$<x,y,z \mid x^z = y, \ y^z = x, \ z^2 = e>$$

(see Exercise 1.7). Prove that the subgroup $H = <x,y>$ is *freely* generated by $\{x,y\}$.

EXERCISE 2.   Compute the order of the group

$$<x,y \mid x^3 yxy^4, y^3 xyx^4> \ .$$

EXERCISE 3.   Find all positive integral solutions of the Diophantine equation

$$1/\ell + 1/m + 1/n = 1 \ .$$

Prove that for each solution (ℓ,m,n), $D(\ell,m,n)' \cong Z \times Z$.

EXERCISE 4.   Prove that $D(\ell,\ell,\ell)'$ is a one-relator group for any $\ell \in N$.

EXERCISE 5.   Show that $(Z_\ell * Z_m)'$ is a free group and write down its rank.

EXERCISE 6.   Perform the computation of Exercise 5 for $(Z_2 \times Z_2) * Z_2$.

EXERCISE 7.   Let M_n be the matrix (of Example 6) determining the Coxeter group S_n . If the $(k,k+1)-$ and $(k+1,k)-$ entries are converted from 3 into 2 $(1 \leq k \leq n-2)$, what is the corresponding Coxeter group?

EXERCISE 8.   Let M be a matrix determining the generalized Coxeter group G . If every row of M contains a non-zero entry on or before the main diagonal, prove that G/G' is finite. Is this condition necessary?

EXERCISE 9   (cf. Exercise 7.5).   Compute the multiplicator of $Z_2 \times Z_2$.   [Time limit: 2 hours .]

EXERCISE 10   (cf. Theorem 8.1).   By finding a presentation for the abelian normal subgroup $H = \langle x^2, y^2, z^2 \rangle$ of

$$J(a,b,c) = \langle x,y,z \mid x^y = y^{b-2}x^{-1}y^{b+2},\ y^z = x^{c-2}y^{-1}z^{c+2},$$
$$z^x = x^{a-2}z^{-1}x^{a+2} \rangle \ ,$$

(where   a,b,c   are non-zero even integers), prove that $J(a,b,c)$ has order $2^8 |abc(abc-1)|$.

§13. A method for presenting subgroups

In this section, we describe a modification of the method of §11 which yields a presentation for the subgroup whose cosets are being enumerated. In fact, if $G = \langle X|R \rangle$ and   H   is the subgroup of   G   generated by $Y = Y(X)$, then the completed enumeration yields a presentation for   H   on the generators   Y .   The modified method thus incorporates both the rewriting process and Tietze transformations described in the previous section.   It therefore provides a more efficient alternative in the case when we are dealing with a particular group and subgroup for which the enumeration terminates.

The basic idea is to keep track of coset representatives rather than just cosets, so that information of the form $ix = j$ is supplemented by $ix = hj$, where   h   is a word in the subgroup generators   Y .   We record this information in our second monitor table only, the others being as before.   The value of the word   h   depends on the source of the information $ix = j$, and is computed according to the following rules.

(i)   If $ix = j$ is a definition, take $h = e$.

(ii)   If $ix = j$ is a consistent bonus coming from the completion of the k-row of the R-table headed by $r = sxt$, we really have

$$ks = h_1 i \, , \quad kt^{-1} = h_2 j \qquad\qquad (1)$$

where h_1, h_2 are words computed inductively from ks and kt^{-1} using the second monitor table. In this case we take $h = h_1^{-1} h_2$.

(iii) If $ix = j$ is a consistent bonus from the Y-table headed by $y = sxt$ and $s = h_1 i$, $t^{-1} = h_2 j$ with h_1, h_2 computed as in (1), we take $h = h_1^{-1} y h_2$.

(iv) If $ix = j$ leads to the collapse $j = k$, with $j < k$ say, we proceed as follows (other types of collapse being treated in a similar way). We have

$$ix = hj \, , \quad ix = h_3 k \, ,$$

where h is computed as in (ii) or (iii) and h_3 is known already. We then substitute $h_3^{-1} h j$ for k throughout, paying heed to the following two points. Firstly, the definition of k is converted into a bonus as before, and secondly, the k-row of the second monitor table needs special attention. A typical entry $kx = h_4 k'$ in this row now yields the information $jx = h^{-1} h_3 h_4 k'$. Now this is either a bonus and is treated as such, or an equation $jx' = h_5 \ell'$ already exists. If $k' \neq \ell'$. it is treated as another collapse and if not, we simply ignore it; the word $h_5^{-1} h^{-1} h_3 h_4$ is a relator of H which will later come out in the wash. As before, we pursue this strategy until all inconsistent information has disappeared.

In the case where the enumeration terminates without any collapse, the relators for H are obtained from those rows of the R-tables which complete without yielding a bonus, namely, the hitherto insignificant animals marked with the symbol ‡ in the foregoing. If the k-row of an R-table is of this type, we have $k = hk$, where $h = h(Y)$ is computed from the second monitor table as in (ii) above. Letting S denote the set of words resulting from all R-rows marked ‡ , $\langle Y | S \rangle$ is a presentation for H . In the case where a collapse has occurred we take S to be the set of words arising in this way from *all* the rows of the R- and Y-tables. This is necessary because a collapse can invalidate our original markings.

Example 1. We resuscitate Example 11.5 and write y = abc for the subgroup generator. The tables are reproduced below, the second monitor table being left blank after the first bonus, namely, 3c = 1 (now interpreted as 3c = yl , according to (ii) above).

a	b	c	a	b⁻¹
1	2	3	1	2	1
2	3	1	2	3	2
3	1	2	3	1	3

b	c	a	b	c⁻¹
1	2	3	1	2	1
2	3	1	2	3	2
3	1	2	3	1	3

c	a	b	c	a⁻¹
1	2	3	1	2	1
2	3	1	2	3	2
3	1	2	3	1	3

a	b	c
1	2	3	1

definition	bonus
1a = 2	
2b = 3	3c = 1 1b = 2 2c = 3
	3a = 1 3b = 1 2a = 3
	1c = 2

	a	b	c	a⁻¹	b⁻¹	c⁻¹
1	2					$y^{-1}3$
2		3		1		
3			y1		2	

The next bonus comes from the completion of the first row of the first table, where we have

1a = 2 ,

2b = 3 => 1ab = 3 ,

3c = yl => 1 abc = yl ,

1a = 2 => 1 abca = y2 .

The modified bonus thus asserts that $y2b^{-1}$ = 1 , that is, 1b = y2 , or $2b^{-1} = y^{-1}1$ (as in (ii) above). In a similar way, we obtain y3 = 2c from the second row of the second table, and $y^2 1 = 3a$ from the third row of the third table. The remaining three bonuses are dealt with in turn in a similar way and our second monitor table finishes up looking like this:

	a	b	c	a^{-1}	b^{-1}	c^{-1}
1	2	$y2$	$y^5 2$	$y^{-2}3$	$y^{-6}3$	$y^{-1}3$
2	$y^5 3$	3	$y3$	1	$y^{-1}1$	$y^{-5}1$
3	$y^2 1$	$y^6 1$	$y1$	$y^{-5}2$	2	$y^{-1}2$

To obtain the relators for H, we first take the second row of the first table, where

$$2 = 2abcab^{-1} = y^5 3bcab^{-1} = y^5 y^6 1cab^{-1} = y^5 y^6 y^5 2ab^{-1}$$

$$= y^5 y^6 y^5 y^5 3b^{-1} = y^5 y^6 y^5 y^5 2 \; .$$

The relator is thus y^{21}. The remaining two rows of type ‡ yield in turn

$$y^6 y^5 y^5 y^6 y^{-1} = y^{21} , \quad y^5 y^5 y^6 y^5 = y^{21} ,$$

for the other relators. We deduce that

$$H = \langle y \,|\, y^{21} \rangle \cong Z_{21} ,$$

so that $|G| = 63$, as in the table of Fibonacci groups given in §9 (p. 76).

Example 2. For purposes of comparison, we now tackle the group

$$G = \langle x,y \mid x^2 yxy^3, y^2 xyx^3 \rangle$$

of Example 12.2, and find a presentation for the subgroup H generated by y. The reader is again counselled to perform the computation himself, using only the definition column of the first monitor table. The enumeration completes without collapse after 8 rows, and the tables are as follows:

x	x	y	x	y	y	y
1	2	3	4	1	1	1	1
2	3	7	3	7	3	4	2
3	7	5	8	8	6	7	3
4	1	2	5	6	7	3	4
5	6	4	2	3	4	2	5
6	4	1	1	2	5	8	6
7	5	6	7	5	8	6	7
8	8	8	6	4	2	5	8

y	y	x	y	x	x	x
1	1	1	2	5	6	4	1
2	5	8	8	6	4	1	2
3	4	2	3	4	1	2	3
4	2	5	6	7	5	6	4
5	8	6	4	2	3	7	5
6	7	3	7	3	7	5	6
7	3	4	1	1	2	3	7
8	6	7	5	8	8	8	8

definition	bonus		x	y	x^{-1}	y^{-1}	y
	$1y=1$	1	2	$y1$	$y^3 4$	$y^{-1}1$	$1 \ddagger 1$
$1x = 2$		2	3	5	1	$y^{-3}4$	
$2x = 3$		3	$y^{-1}7$	4	2	$y^{-3}7$	
$3y = 4$	$4x=1$ $4y=2$	4	$y^{-3}1$	$y^3 2$	$y^6 6$	3	
$2y = 5$		5	6	8	$y^{-3}7$	2	
$5x = 6$	$6x=4$	6	$y^{-6}4$	7	5	$y^{-1}8$	
$6y = 7$	$7y=3$ $7x=5$ $3x=7$	7	$y^3 5$	$y^3 3$	$y3$	6	
$5y = 8$	$8y=6$ $8x=8$	8	$y^8 8$	$y6$	$y^{-8}8$	5	

Using the second monitor table, the eight rows marked $\ddagger$ yield the following relators for h respectively:

$$ey^{-1}y^3 y^{-1}y^3 ey^3$$

$$y^{-1}y^3 ey^8 yey^3 \qquad yyeeey^{-6}y^{-3}$$

$$y^{-6}y^{-3}yeeey \qquad ey^3 y^{-1}y^3 y^{-1}y^3 e$$

$$y^3 eey^3 eye \qquad yey^3 ey^8 y^8 y^8$$

$$y^8 y^8 yy^{-6}y^3 ee$$

We see that $H = \langle y \mid y^7 \rangle \cong Z_7$, and this is an alternative proof that $|G| = 56$.

Example 3. To give an application of the method when the sub-

group involved is non-cyclic, consider the subgroup H generated
by x = ab, y = ac in the group

$$G = F(3,4) = \langle a,b,c,d \mid abc = d,\ bcd = a,\ cda = b,\ dab = c\rangle .$$

The tables are completed in the usual way and turn out to be as
follows:

a	b	c	d^{-1}
1	2	1	3	1
2	4	2	1	2
3	1	3	4	3
4	3	4	2	4

b	c	d	a^{-1}
1	3	4	2	1
2	1	3	4	2
3	4	2	1	3
4	2	1	3	4

c	d	a	b^{-1}
1	3	4	3	1
2	1	3	1	2
3	4	2	4	3
4	2	1	2	4

d	a	b	c^{-1}
1	3	1	3	1
2	1	2	1	2
3	4	3	4	3
4	2	4	2	4

definition	bonus			
1a = 2	2b = 1	2c = 1	2d = 1	
1c = 3	1d = 3	3a = 1	1b = 3	
2a = 4	4b = 2	3d = 4	4a = 3	
	3c = 4	4d = 2	3b = 4	
	4c = 2			

y

a	c
1	2	1

x

a	b
1	2	1

	a	b	c	d	a^{-1}	b^{-1}	c^{-1}	d^{-1}
1	2	$x^{-1}y3$	3	$x3$	$x^{-1}yx3$	$x^{-1}2$	$y^{-1}2$	$xy^{-1}2$
2	4	$x1$	$y1$	$yx^{-1}1$	1	$yxy^{-1}4$	$yx^{-1}yxy^{-1}4$	$x^{-1}y4$
3	$x^{-1}y^{-1}x1$	$y^{-1}x4$	4	$x^{-1}4$	$y^{-1}4$	$y^{-1}x1$	1	$x^{-1}1$
4	$y3$	$yx^{-1}y^{-1}2$	$yx^{-1}y^{-1}xy^{-1}2$	$y^{-1}x2$	2	$x^{-1}y3$	3	$x3$

We deduce in turn the following relations for H from the five
rows marked ‡ :

$$(y)(y^{-1}x)(yx^{-1}y^{-1}xy^{-1})(x^{-1}y) = e$$

$$(y^{-1}x)(yx^{-1}y^{-1}xy^{-1})(yx^{-1})(x^{-1}yx) = e$$

$$(yx^{-1}y^{-1})(y)(x)(y^{-1}) = e$$

$$(e)(y^{-1}x)(e)(x^{-1}y) = e$$

$$(y^{-1}x)(e)(yx^{-1}y^{-1})(yx^{-1}yxy^{-1}) = e$$

As the last of these is a consequence of the first two and the other two are trivial, we have

$$H = \langle x,y \mid [y,x] = [x,y^{-1}] = [x^{-1},y]\rangle$$

Note that $H^{ab} \cong Z \times Z$, so that $F(3,4)$ is infinite (as claimed in the table of §9).

Example 4. To illustrate the phenomenon of coset collapse, we take as our final example the subgroup $H = \langle x \rangle$ of $F(2,5)$ (see Example 11.4):

$$G = \langle x,a,b,c,d \mid xa = b, \ ab = c, \ bc = d, \ cd = x, \ dx = a\rangle \quad .$$

Since the tables actually collpase down to one row, this example also shows how the method can be used to find a presentation of a group on a new set of generators (cf. Theorem 4.4). The tables are reproduced below, with the second monitor table suitably modified.

x	a	b^{-1}
1	1 3	1
2	3	2
3	2	3

a	b	c^{-1}
1	3 2	1
2	1 3	2
3		3

b	c	d^{-1}
1	3 2	1
2	1	2
3	2 3	3

c	d	x^{-1}
1	2 1	1
2	3 3	2
3		3

d	x	a^{-1}
1	2 3	1
2	1 1	2
3	3	3

definition	bonus		x
	1x = 1		1	1
1c = 2	2d = 1 2a = 1		
1a = 3	1b = 3 3b = 2 2c = 3 3d = 3	
	2x = 3 1d = 2 $\underline{3c = 2}$	

	x	a	b	c	d	x^{-1}	a^{-1}	b^{-1}	c^{-1}	d^{-1}
1	x1	3	x3	2	$x^{-6}2$	$x^{-1}1$	$x^{-2}2$			$x^{-1}2$
2	$x^{6}3$	$x^{2}1$		$x^{3}3$	x1			3	1	$x^{6}1$
3			2		$x^{3}3$	$x^{-6}2$	1	$x^{-1}1$	$x^{-3}2$	$x^{-3}3$

125

Now the row marked $\ddagger$ yields

$$x3c = (1b)c = 1d = x^{-6}2 \ ,$$

and since we already have $1 = 2c^{-1}$, we deduce $3 = x^{-7}1$. We then find that

$$2 = 3b = x^{-7}1b = x^{-6}3 = x^{-13}1 \ ,$$

and the collapse is complete. The first half of the first row of the modified monitor table now takes the form:

	x	a	b	c	d
1	x1	$x^{-7}1$	$x^{-6}1$	$x^{-13}1$	$x^{-19}1$

Thus we find that

$$x^{-6}1 = 1b = 1xa = x1a = x1a = xx^{-7}1 \ ,$$

yielding the empty word. The remaining four relators of G yield

$$x^{-13}1 = x^{-7}x^{-6}1, \quad x^{-19}1 = x^{-6}x^{-13}1 \ ,$$
$$x1 = x^{-13}x^{-19}1, \quad x^{-7}1 = x^{-19}x1 \ ,$$

and we deduce that

$$G = H = \langle x \mid x^{33}, x^{11} \rangle = \langle x \mid x^{11} \rangle \cong Z_{11} \ ,$$

as before.

We conclude this section by proving that the method does in fact yield a presentation for the subgroup in question. We adopt the notation and assumptions of the last part of §11 (p.103). Thus the definitions inductively determine a Schreier transversal for $\bar{R}\langle Y \rangle$ in F , while the bonuses yield a set

$$B = \{r_i h_i \mid 1 \le i \le m\}, \ m = (n-1)g + 1, \ r_i \in \bar{R}, \ h_i \in \langle Y \rangle \ ,$$

of free generators for $\bar{R}<Y>$ as in Exercise 11.11 and §12. Thus the $h_i = h_i(Y)$ generate $<Y>$ (though not necessarily freely), and are precisely the words in Y figuring in the first half of the modified second monitor table. Now for each $r \in R$ and $1 \le k \le g$, let

$$w_{r,k} = w_{r,k}(r_1 h_1, \ldots, r_m h_m)$$

be the word in B obtained by letting r act letterwise on u_k , that is, $u_k r = w_{r,k} u_k$. It follows from Theorem 12.1 that $\bar{R}$ is the normal closure of the set

$$\hat{S} = \{w_{r,k} \mid r \in R, \ 1 \le k \le g\}$$

in $\bar{R}<Y>$. (The letterwise action simply corresponds to the re-writing process of §12.) Furthermore, the words $w_{r,k} = w_{r,k}(h_1, \ldots, h_m)$ obtained by deleting the r_i's from $w_{r,k}$ is precisely the set S of alleged relators for H .

Since $\bar{R}<Y>$ is free on B , there is a unique homomorphism

$$\left.\begin{array}{rcl} \theta: \bar{R}<Y> & \to & <Y> \\[2mm] r_i h_i & \mapsto & h_i \end{array}\right\} \ .$$

Now H is the factor group of $\bar{R}<Y>$ by the normal closure of $\hat{S}$, and since the h_i generate $<Y>$, θ is onto. It follows from Exercise 1.3 that H is isomorphic to the factor group of $<Y>$ by the normal closure N of $\hat{S}\theta = S$. (This corresponds to the Tietze transformations of §12.) Since $<Y>$ may not be *freely* generated by Y , we must perform one final step. There is a free presentation $\nu: F(Y) \twoheadrightarrow <Y>$ fixing the generators (by san), and regarding S as a subset of $F(Y)$, we have $\bar{S}\nu = \overline{S\nu} = N$ by Exercise 1.3 again. It follows that

$$H \cong <Y>/N \cong F(Y)/\bar{S} = <Y|S> \ ,$$

as required.

EXERCISE 1 (see Example 3). Use Tietze transformations to show that

$$F(3,4) = \langle a,b,c,d \mid abc = d, \ bcd = a, \ cda = b, \ dab = c\rangle$$
$$\cong G \quad = \langle a,b,c \mid abc = c^{-1}b^{-1}a = c^{-1}ba^{-1} = cb^{-1}a^{-1}\rangle \quad ,$$

and deduce that $a^2 = b^2 = c^2 \in Z(G)$. Use this to write the commutators $[a,b],[a,c],[b,c]$ as words in $x = ab$, $y = ac$ and deduce that $H/G' \cong Z_2 \times Z_2$. Use the method of the previous section to find a presentation for G' .

EXERCISE 2. Use the method of this section to show that the subgroup H generated by x^2,y^2,xy in $F = \langle x,y \mid \ \rangle$ has index 2 and is free on these generators.

EXERCISE 3. Find a presentation of the subgroup $\langle x\rangle$ of

$$\langle x,y \mid x^3yxy^4,y^3xyx^4\rangle \quad .$$

EXERCISE 4. Present the subgroup $\langle ab,bc,ca\rangle$ of

$$G = \langle a,b,c \mid abc = b, \ bca = c, \ cab = a\rangle \quad .$$

Can you identify G ?

EXERCISE 5. Check that the subgroup $\langle x^2,y^2,z^2\rangle$ of

$$\langle x,y,z \mid x^{-1}yx^{-1}y^3,y^{-1}zy^{-1}z^3,z^{-1}xz^{-1}x^3\rangle$$

is normal and abelian, and find its order and index.

EXERCISE 6. Find a presentation for the subgroup $\langle x^2,y^2,z^2\rangle$ of

$$\langle x,y,z \mid x^{-1}y^{-1}xy^3,y^{-1}z^{-1}yz^3,z^{-1}x^{-1}zx^3\rangle \quad .$$

EXERCISE 7.   Can you name the groups of the last two exercises?
Are they isomorphic?

EXERCISE 8.   Identify the subgroup of

$$\langle a,b,c \mid abcab^{-1}, bcabc^{-1}, cabca^{-1} \rangle$$

generated by $x = abc,\ y = ab$.

5 · The triangle groups

What immortal hand or eye

Could frame thy fearful symmetry? (Blake: The tiger)

Suppose we wish to tesselate a space with copies of a triangle ABC :

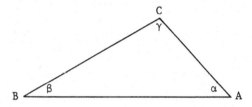

If a,b,c denote reflection in BC,CA,AB respectively, then

$$a^2 = b^2 = c^2 = e \quad . \tag{1}$$

Furthermore, bc consists of a counterclockwise rotation about A through an angle 2α . Thus, to have any hope of success, we must insist that 2π is an integer multiple of 2α , and similarly of 2β and 2γ . Taking

$$\alpha = \pi/m, \ \beta = \pi/n, \ \gamma = \pi/\ell, \ 2 \leq \ell,m,n \in Z \ ,$$

it follows that

$$(ab)^\ell = (bc)^m = (ca)^n = e \quad . \tag{2}$$

The group of transformations generated by a,b,c is thus a homomorphic image of the group $\Delta = \Delta(\ell,m,n)$ of Definition 12.1. Indeed, under suitable conditions, these groups turn out to be the same, so that the triangles involved in the tesselation are in one-to-one correspondence with the elements of the group Δ . It

is clear from the relations (1),(2) that the words of length two
in Δ comprise a subgroup of index 2 generated by ab,bc .
This is called the group of 'pure rotations', and is isomorphic
to the von Dyck group $D(\ell,m,n)$ (Example 12.5).

To decide whether $D(\ell,m,n)$ is finite or not, we take a closer
look at the space in which our triangle is sitting. The nature of
this space depends critically on the angle-sum of the triangle;
specifically, three cases arise according as $1/\ell + 1/m + 1/n$ is
equal to, greater than, or less than unity. We take each of these
in turn, observing that $D(\ell,m,n)$ is a finite group if and only
if the corresponding space is compact.

§14. The Euclidean case

We begin with the case $1/\ell + 1/m + 1/n = 1$, as this is
traditionally the most familiar of the three. Assuming without
loss of generality that $2 \leq \ell \leq m \leq n$, our equation has only
three solutions (Exercise 12.3):

$(3,3,3)$, $(2,4,4)$, $(2,3,6)$,

and we take each of these in turn.

Consider the tesselation of R^2 given by the equilateral
triangle, of which a piece is shown in Fig.1 (p.132). Our original
triangle ABC is labelled e here, and its images under a few of
the elements of Δ are labelled accordingly. Note that alternate
triangles are labelled with words of even length, corresponding to
members of $D(3,3,3)$.

In order to prove that $\Delta(3,3,3)$ is an infinite group, it
will suffice to show that *every* triangle acquires a label. This
is a consequence of two simple topological facts, both of which
are expressed in terms of the following notion. Two triangles
are called *adjacent* if they have an edge in common, and are called
connected if there is a finite sequence of triangles containing
these two such that each is adjacent to its successor. Then we
observe that:

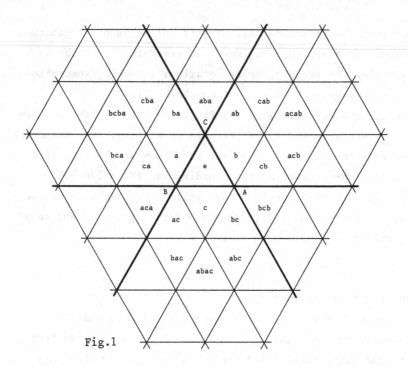

Fig.1

(i) any reflection is continuous, whence the members of Δ
preserve adjacency of triangles, and

(ii) the dual graph (tesselation of R^2 by regular hexagons)
is connected, whence every triangle is connected to ABC .

Now suppose that there is an unlabelled triangle. Since this
is connected to ABC (by (ii)), we can find en route adjacent
triangles T,T' such that T has a label, w say, while T' is
unlabelled. Now by construction, Tw^{-1} = ABC , and is adjacent to
$T'w^{-1}$ (by (i)), so that $T'w^{-1}$ = (ABC)x , where x is one of
a,b,c . Our construction now implies that T' bears the label
xw , which is a contradiction. Thus, Δ(3,3,3) (and hence
also D(3,3,3)) is infinite, it is left as an exercise to show
that its elements are actually in one-to-one correspondence with
the triangles in the tesselation.

For the sake of variety, we tackle the group Δ(4,4,2) in a
slightly different way. Suppose the plane is tesselated by
squares, with *edges* labelled as in the following diagram, where
the heavily-marked square is understood to repeat indefinitely up
and down and to left and right.

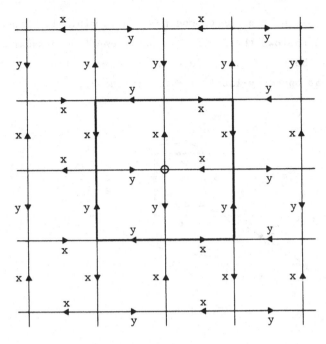

Fig.2

Since the four edges at any vertex bear the labels $x^{\pm 1}, y^{\pm 1}$ in some order, both x and y induce permutations of the vertex set (thought of as the Gaussian integers Γ in C with origin O). We thus have a homomorphism: $\langle x,y| \rangle \to S_\Gamma$, and $x^4, y^4, (xy)^2$ clearly belong to its kernel. There results a homomorphism

$$\rho: D(4,4,2) \to S_\Gamma$$

and since $(x^{-1}y)\rho$ sends $2k$ to $2(k+1)$ for all $k \in Z$, Im ρ (and hence $D(4,4,2)$) must be infinite. We shall now prove that ρ is one-to-one, so that Fig.2 is nothing other than the graph (Cayley diagram) of $D(4,4,2)$ with respect to $\{x,y\}$.

To see this, observe that any reduced word $w = w(x,y) \in D(4,4,2)$ defines (via ρ) a path P in the diagram starting at O . It is clear that if $w\rho = e$, then this path is actually a *loop* at O , and as such encloses a finite number (k, say) of square tiles. To show that ρ is faithful, we assume (for a contradiction) that $w \neq e$ is a reduced word with $w\rho = e$, such that the number k

of tiles enclosed by the corresponding path is minimal. Since w
is reduced, it follows that $k \geq 1$, and we can find a square inside P whose boundary has an edge z in common with P . If
this square has boundary-label r = zt (reduced), then

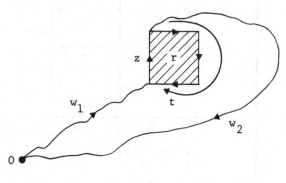

Fig.3

$$z = x^{\pm 1} \text{ or } y^{\pm 1} \text{ , and } r = x^{\pm 4}, y^{\pm 4}, (xy)^{\pm 2} \text{ or } (yx)^{\pm 2} .$$

According to Fig.3, we have

$$w = w_1 z w_2 = w_1 r t^{-1} w_2$$

$$= (w_1 r w_1^{-1}) w_1 t^{-1} w_2 .$$

Now $w_1 t^{-1} w_2$ is a word in $D(4,4,2)$ for which the corresponding
path is a loop at O containing only k-1 squares, whence
$w_1 t^{-1} w_2 = e$ by minimality. Since $w_1 r w_1^{-1}$ is also equal to e
in $D(4,4,2)$, so is w , and this is the required contradiction.
    In the final case, where $\{\ell, m, n\} = \{2,3,6\}$, we content our-
selves with drawing the pictures. Thus, Fig.4 shows the
tesselation obtained from $\Delta(2,3,6)$, and Fig.5 illustrates the
graph of $D(3,6,2)$.

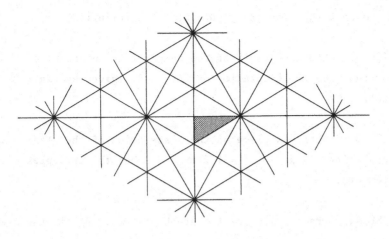

Fig.4

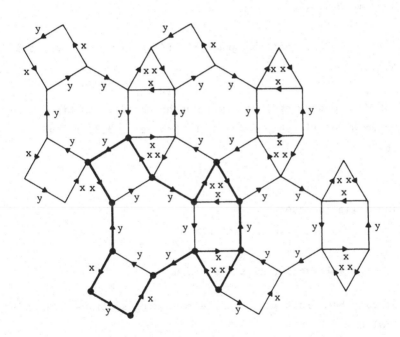

Fig.5

EXERCISE 1. Use the permutations

$$\xi = \ldots(012)(345)\ldots, \quad \eta = \ldots(123)(456)\ldots$$

of Z to show that the group D(3,3,3) is infinite.

EXERCISE 2.   Use the result of the computation in Exercise 12.3
to show that the representation of $\Delta(3,3,3)$ given by Fig.1 is
faithful.

EXERCISE 3.   Sketch a piece of the tesselation of the plane
obtained using the group $\Delta(4,4,2)$, and label the triangles with
its elements.

EXERCISE 4.   Draw a piece of the graph of D(3,3,3) in the plane.

EXERCISE 5.   Prove that all the triangles in Fig.4 receive
labels from $\Delta(2,3,6)$.

EXERCISE 6.   Prove that the graph of D(3,6,2) is as depicted
in Fig.5.

EXERCISE 7.   Observing that Fig.1 can be embedded in Fig.4, can
you derive a relationship between the groups $\Delta(3,3,3)$ and
$\Delta(2,3,6)$?

§15. The elliptic case
    All the groups $D(\ell,m,n)$ for which

$$1/\ell + 1/m + 1/n > 1 , \quad 2 \leq \ell \leq m \leq n ,$$

have already been identified in the foregoing sections.   The list
is as follows:

(ℓ,m,n)	group	order	figure
(2,2,n)	D_n	2n	n-gon
(2,3,3)	A_4	12	tetrahedron
(2,3,4)	S_4	24	octahedron
(2,3,5)	A_5	60	icosahedron

Note that in each case, the group is the symmetry group of a regular figure in R^3 , and that the group order is equal to $2(1/\ell + 1/m + 1/n - 1)^{-1}$.

To achieve a tesselation in this case, we need a space in which triangles have angle-sum exceeding π . The obvious candidate is the surface of the unit ball in R^3 , with great circles playing the role of lines. As we shall see, tesselations exist in every case, and the triangles appearing are in one-to-one correspondence with the elements of the group Δ . To see this, note that the area of such a triangle is equal to the amount by which its angle-sum exceeds π (the Gauss-Bonnet theorem). Thus, all the triangles in the (ℓ,m,n)-tesselation have area equal to $(\pi/\ell + \pi/m + \pi/n - \pi)$, and they partition a surface of area 4π . Hence, the number of triangles involved is just

$$4\pi(\pi/\ell + \pi/m + \pi/n - \pi)^{-1} = 4(1/\ell + 1/m + 1/n - 1)^{-1}$$
$$= 2|D(\ell,m,n)|$$
$$= |\Delta(\ell,m,n)| \; ,$$

from the above table.

Starting with the dihedral case $\ell = m = 2$, take any triangle with one vertex at the North Pole and the other two on the equator distance π/n apart. Reflections in the two longitudinal sides partition the northern hemisphere into $2n$ triangular regions congruent to the original one, which thus has angle π/n at the pole, as well as $\pi/2, \pi/2$ on the equator. Reflection in the equator duplicates this picture in the southern hemisphere, and we achieve the desired tesselation. The case $n = 6$ is illustrated in Fig.1 below.

Turning to the tetrahedral case, we begin by inscribing a cube in the unit sphere in such a way that two adjacent vertices lie on the Greenwich meridian at points equidistant from the equator. These vertices are labelled A and C in the projection of Fig.2. A little thought shows that there are exactly six great circles containing more than two vertices of the cube, given by projecting the edges of the cube from the centre onto the surface of the

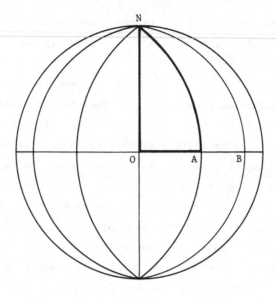

Fig.1

sphere. There results a division of the surface into 24 triangles, of which a typical one is NAB . Two great circles meet
orthogonally at N , so the angle here is π/2 , and since three
great circles meet at each of A and B , the angles here are
both equal to π/3 .

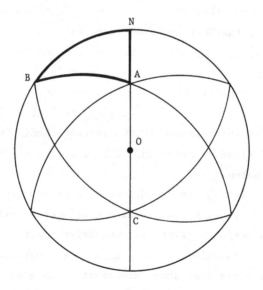

Fig.2

Now consider the angle-bisector at N in the triangle ANB.
It clearly lies on the line of longitude 45° W, and since ANB
is isosceles, it meets BA orthogonally. Denoting this point of
intersection by D, we have

$$N\hat{D}A = \pi/2 , \quad D\hat{A}N = \pi/3 , \quad A\hat{N}D = \pi/4 ,$$

and we land happily in the octahedral case. Drawing in the
equator and the great circle containing longitude 45° E, each
of our original 24 triangles is bisected once, and we obtain
the tesselation illustrated in Fig.3. Note that the six points
where these three extra circles meet comprise the vertices of a
regular octahedron.

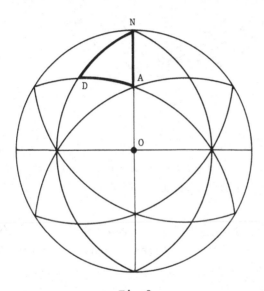

Fig.3

Finally, we approach the case $(\ell,m,n) = (2,3,5)$ by imagining
a regular icosahedron whose vertices lie on the unit sphere.
Central projection maps the 30 edges into 15 pairs of antipodal great circle arcs. These 15 circles meet:

    in fives at the vertices,
    in threes at the centres of the faces, and
    in twos at the mid-points of the edges.

There results a barycentric subdivision of each of the 20 faces into 6 congruent triangles, with angles $\pi/5, \pi/3, \pi/2$, as required. Fig.4 illustrates the situation, with heavy dots marking the visible vertices of the icosahedron.

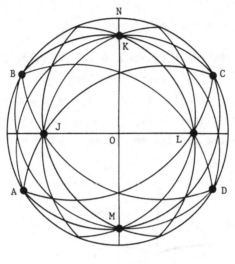

Fig.4

EXERCISE 1.   Taking the scale ON = 1 in Fig.1, prove that OA = $\frac{1}{2}$, OB = $\sqrt{3}/2$, and deduce that this figure can be constructed with ruler and compasses.

EXERCISE 2.   Again taking ON = 1, prove that OA = $1/\sqrt{3}$ in Fig.2, and derive a ruler-and-compass construction for this figure.

EXERCISE 3.   Show that the tesselation of the sphere in Fig.2 is the same as that obtained by central projection of the edges of an inscribed regular tetrahedron.

EXERCISE 4.   With reference to Fig.4, note that pairs of vertices of the icosahedron are of three types:

   adjacent, antipodal, other,

whose distances apart are:

AB = CD = JL = p , say,

AC = BD = 2 ,

BC = AD = KM = q , say,

respectively. Prove that

$$q = p \, \frac{1 + \sqrt{5}}{2} \, , \quad p = \sqrt{2 - 2/\sqrt{5}} \, ,$$

and derive a ruler-and-compass construction for this figure.

EXERCISE 5.   Does the method used in Figs.1-4 to illustrate the tesselations correspond to any kind of natural projection?

EXERCISE 6.   Prove that

$$\Delta(3,3,2) \cong S_4 \cong D(2,3,4) \, ,$$

and that in every other elliptic case,

$$\Delta(\ell,m,n) \cong D(\ell,m,n) \times Z_2 \, .$$

§16. The hyperbolic case

In order to study the groups $\Delta(\ell,m,n)$ for which $1/\ell + 1/m + 1/n < 1$, we need a space where the angle-sum of a triangle is less than π . An ideal example consists of an open disc in R^2 , in which the lines are just diameters of the disc together with arcs of circles orthogonal to its boundary. While the notion of reflection retains its usual meaning for genuine lines (diameters), it is interpreted as inversion when the line in question is a circular arc. Before going on to discuss the relevant properties of inversion in a circle, we demonstrate that this space contains triangles of the required type.

Given positive integers ℓ, m, n with $1/\ell + 1/m + 1/n < 1$, we begin by forming a triangle whose angles are $\pi/\ell, \pi/m, \pi/n$, and end by drawing a circle Ω orthogonal to each of its sides (two straight, one curved). The construction (which, incidentally, is *not* of the ruler-and-compass variety) is illustrated in Fig.1.

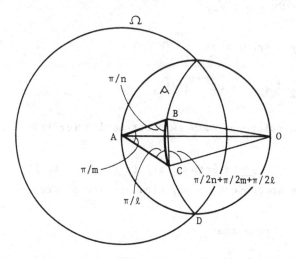

Fig.1

Let ABC be a Euclidean triangle with angles at A,B,C respectively equal to

$$\pi/m, \quad \pi/2 + \pi/2n - \pi/2\ell - \pi/2m, \quad \pi/2 + \pi/2\ell - \pi/2n - \pi/2m \; .$$

Next, let OBC be the isosceles Euclidean triangle with base BC and base angle $\pi/2n + \pi/2m + \pi/2\ell$. The arc BC of the circle centre O then meets AB and AC in angles

$$(\pi/2 + \pi/2n - \pi/2\ell - \pi/2m) + (\pi/2n + \pi/2m + \pi/2\ell) - \pi/2 = \pi/n \; ,$$

$$(\pi/2 + \pi/2\ell - \pi/2n - \pi/2m) + (\pi/2n + \pi/2m + \pi/2\ell) - \pi/2 = \pi/\ell \; ,$$

respectively, and the non-Euclidean triangle ABC has the required angles. Now let D be a point where the circle $\triangle$ centre O radius OB meets the circle diameter AO. Since the angle ADO is a right angle, the circle Ω centre A radius AD is

142

orthogonal to all the sides of our non-Euclidean triangle ABC .

Letting b,c denote reflection in AC,AB respectively, and a inversion in A , it is clear that b and c map the interior of Ω bijectively onto itself, and that $(bc)^m$ fixes each of its points. The corresponding facts involving the inversion a are also true and, at the expense of slight digression, we proceed to establish them now.

Let Γ denote the unit circle centre the origin O in R^2 , and γ the operation on $R^2\backslash\{O\}$ of inversion with respect to Γ . If (r,θ) are the polar coordinates of a typical point of $R^2\backslash\{O\}$, then its image under γ is the point $(1/r,\theta)$, whence γ constitutes an involution on this space. It is easy to prove that γ maps circles to circles, and that its set of fixed points is just Γ . Furthermore, if we define an 0-circle (0-line) to be the intersection with $R^2\backslash\{O\}$ of a circle (line) through the origin, then γ interchanges lines with 0-circles and fixes 0-lines. We need one more crucial property of γ , namely that *it preserves angles* (see Exercise 2). It is clear that all these properties carry over to the operation of inversion in an arbitrary circle, simply by adjusting the coordinate system.

It follows from what has been said that γ preserves circles orthogonal to Γ , so that the inversion a in A (Fig.1) maps the inside of Ω onto itself. Our main objective is now to demonstrate that $(ca)^n$ is the identity, and this is done by means of the following lemma.

Lemma 1. *Let* P *and* Q *be points (not necessarily distinct) of* $R^2\backslash\{O\}$. *Then,*

(i) if P *and* Q *are mirror images in an 0-line* Λ *, so are* Pγ *and* Qγ ,

(ii) if P *and* Q *are inverse with respect to an 0-circle* A, *then* Pγ *and* Qγ *are mirror images in the line* $A\gamma$.

Proof. Consider the following diagrams.

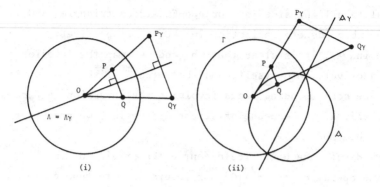

Fig.2

The first assertion is obvious from Fig.2(i), since Λ is an
altitude of the triangle OPQ . For the second part, note that
PQ meets Λ in a right angle, so the lines Aγ and PγQγ are
perpendicular since γ preserves angles. Now if Ƃ is any
circle through P and Q , it must be fixed by inversion in Λ ,
whence Λ and Ƃ are orthogonal. It follows that Ƃγ is
orthogonal to Aγ , whence its centre lies on this line. Since
Pγ and Qγ lie on Ƃγ , they must be equidistant from Aγ , and
this completes the proof.

Now let Λ be a line meeting a circle A in an angle π/n ,
where n is an integer greater than 1 . If Λ and A meet in
points O and A , choose coordinates so that Γ is the circle
centre O radius OA , as in Fig.3.

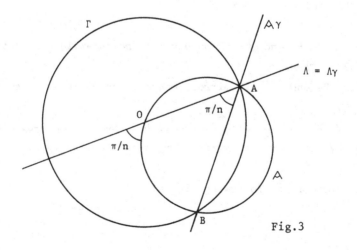

Fig.3

If Γ meets Λ again in B , then $\Lambda\gamma$ is just the line through AB , and $O\hat{A}B$ is just π/n since γ preserves angles. Let λ,μ denote reflection in $\Lambda,\Lambda\gamma$, respectively, and let α be inversion in Λ . We call a point of R^2 *pathological* if it either coincides with O , or can be mapped to the centre of Λ by a word in λ,α . Taking any non-pathological point P , let $Q = P\lambda$ and apply Lemma 1(i) to obtain that $P\gamma$ and $P\lambda\gamma$ are mirror images in Λ , that is, the transformations $\gamma\lambda$ and $\lambda\gamma$ agree on non-pathological points. Similarly, we apply Lemma 1(ii) to P and $P\alpha$, and deduce that $P\gamma$ and $P\alpha\gamma$ are interchanged by μ . It follows that, on non-pathological points,

$$\lambda = \gamma\lambda\gamma \ , \quad \alpha = \gamma\mu\gamma \ .$$

Now we already know (from the introduction to this chapter) that $(\lambda\mu)^n$ is the identity on R^2 , and it follows that $(\lambda\alpha)^n$ fixes every point on which it can be defined. Returning to our original situation (Fig.1), this means that $(ca)^n$ fixes the interior of Ω pointwise. Since the same is true of $(ab)^\ell$, we have shown that the transformations a,b,c again generate a homomorphic image of $\Delta(\ell,m,n)$.

That the images of the triangle ABC under words in a,b,c do in fact cover the inside of Ω , a more delicate analysis is required than that carried out in the Euclidean case (§14). Luckily however, we are now in a position to prove that $\Delta(\ell,m,n)$ is infinite without recourse to this result. To see this, let L be the set of images of ABC under words in a,b,c , each bearing its appropriate label. Further, let R be the region formed by intersecting the closed disc bounded by Λ with the inside of Ω . Defining

$$d = \begin{cases} (bc)^{m/2} & , \quad m \text{ even} , \\ (bc)^{(m-1)/2}b & , \quad m \text{ odd} , \end{cases}$$

we see that Rd is a copy of R , obtained by rotation through π about A (m even), or reflection in A (m odd). If k denotes the diameter of Ω parallel to ED , then R and ABC

lie in one of the semicircles defined by D , while Rd lies in the other. Hence, Rd is disjoint from both R and ABC . Now suppose that R contains r labelled triangles from L ; then so do both Rd and Ra . Since Ra contains both Rd and also ABC , it follows that r+1 ≤ r , and we deduce that r cannot be finite.

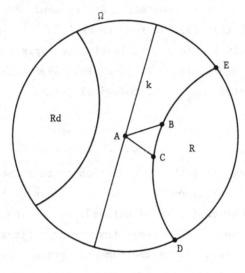

Fig.4

EXERCISE 1.   Prove that inversion maps circles to circles, interchanges lines with 0-circles, and fixes 0-lines.

EXERCISE 2.   Show that inversion preserves tangency of 0-circles and 0-lines, and deduce that it also preserves angles.

6 · Extensions of groups

> ... Though with patience He stands waiting,
> with exactness grinds He all.
>
> (Longfellow's translation of von Logau: Retribution)

The main purpose of this chapter is to give a proof of the
celebrated theorem of Golod and Šafarevič, which gives an accurate
lower bound for the minimal number $r'(G)$ of relations needed to
define a finite p-group G minimally generated by $d(G)$ elements.
The naive bound $r'(G) \geq d(G)$ of Theorem 6.7 is relegated to the
humble role of a lemma, to be invoked almost unconsciously in the
penultimate line of the proof. The proof we give is due to P.
Roquette and is extremely elegant, modulo the rather technical
machinery needed to begin it. We shall need the notions of a G-
module A , and of the cohomology groups $H^n(G,A)$, $n \in N_o$. If
the field k of p-elements is made into a G-module in a trivial
way, it turns out that $H^1(G,k)$ is a vector space over k of
dimension $d(G)$, while the dimension $r(G)$ of $H^2(G,k)$ is at
most $r'(G)$. The last fact is proved in Theorem 20.3 using an
argument based on the presentation theory of group extensions.
Thus we begin with an account of the classical theory of group
extensions, and then proceed to establish the connection with
group cohomology. The only remaining preliminaries are a localization process and some basic facts about finite p-groups. We
conclude by shedding a little light on the unsolved problem of
classifying those finite p-groups for which $r(G)$ and $r'(G)$ are
equal.

§17. Extension theory

The aim of extension theory is to classify those groups K
having a normal subgroup A when the groups A and $G = K/A$
have been specified (up to isomorphism) in advance. Now any

$x \in K$ gives rise to an automorphism

$$\left. \begin{array}{l} \alpha_x \colon A \to A \\ \quad a \mapsto x^{-1}ax = a^x \end{array} \right\}$$

of A and, combining these, we obtain a homomorphism

$$\left. \begin{array}{l} \alpha \colon K \to \text{Aut } A \\ \quad x \mapsto \alpha_x \end{array} \right\} .$$

Now any homomorphism $\beta \colon G \to \text{Aut } A$ yields a homomorphism

$$K \xrightarrow{\text{nat}} G \xrightarrow{\beta} \text{Aut } A \;\;,$$

and α will be of this form if and only if $A \le \text{Ker } \alpha$. Since $\text{Ker } \alpha = C_K(A)$, this is in turn equivalent to the condition that A be abelian.

To avoid complications, we will assume from the outset that A is abelian and write it additively. The homomorphism β induced by α takes the form

$$\left. \begin{array}{l} \beta \colon G \to \text{Aut } A \\ \quad Ax \mapsto \left\{ \begin{array}{l} A \to A \\ a \mapsto a^x \end{array} \right. \end{array} \right\} ,$$

and corresponds in a natural way to an action β' of G on A given by

$$\left. \begin{array}{l} \beta' \colon A \times G \to A \\ \quad (a,g) \mapsto ag \end{array} \right\} ,$$

where $ag = a^x$ when $g = Ax$. It is a simple matter to check that this makes A into a G-module in the following sense.

Definition 1. Given a (multiplicative) group G , a G-*module* is an (additive) abelian group A together with a mapping

$$\left. \begin{array}{l} A \times G \to A \\ (a,x) \mapsto ax \end{array} \right\}$$

which obeys the following three laws:

$$(a+b)g = ag + bg , \quad a(gh) = (ag)h , \quad ae = a , \tag{1}$$

for all $a,b \in A$, $g,h \in G$.

In terms of these ideas, we modify our original problem as follows. We shall study those groups K having a prescribed abelian
normal subgroup, with prescribed factor group, and such that conjugation within K induces a prescribed action of the latter on the
former.

Definition 2. An *extension* of a group G by a G-module A is a
diagram

$$K: \quad A \overset{\iota}{\to} K \overset{\nu}{\to} G ,$$

where:
(i) K is a group and ι,ν are group homomorphisms,
(ii) ι is one-to-one, $\mathrm{Im}\ \iota = \mathrm{Ker}\ \nu$ and ν is onto,
(iii) conjugation in K induces the prescribed G-action on A .

Remark 1. Notice that

$$A \cong \mathrm{Im}\ \iota = \mathrm{Ker}\ \nu \lhd K , \quad K/\mathrm{Ker}\ \nu \cong \mathrm{Im}\ \nu = G ,$$

so that K has a normal subgroup isomorphic to A , with factor
group isomorphic to G .

Remark 2. We often write K in the form

$$0 \to A \overset{\iota}{\to} K \overset{\nu}{\to} G \to 1 , \tag{2}$$

where $0,1$ are trivial groups, and refer to this as a *short exact sequence* of groups. Part (ii) of the definition thus asserts that the image of each homomorphism of (2) is the kernel of its successor.

Remark 3.   Part (iii) of the definition asserts that if $x \in G$ and $k \in \nu^{-1}(x)$, then for all $a \in A$,

$$(ax)\iota = k^{-1}(a\iota)k \quad . \tag{3}$$

The right hand side is independent of the choice of k since A is abelian.

Example 1.   Given a group G and a G-module A , consider the Cartesian product $S = G \times A$ imbued with the following binary operation:

$$(x,a)(y,b) = (xy,ay+b) \quad , \tag{4}$$

for all $x,y \in G$, $a,b \in A$. We compute that for all $(x,a),(y,b),(z,c) \in G \times A$,

$$((x,a)(y,b))(z,c) = ((xy)z,(ay+b)z+c) \quad ,$$

$$(x,a)((y,b)(z,c)) = (x(yz),a(yz)+(bz+c)) \quad ,$$

and these are equal because of (1) above. This also guarantees that

$$(x,a)(e,0) = (xe,ae+0) = (x,a) \quad ,$$

and furthermore,

$$(x,a)(x^{-1},-ax^{-1}) = (xx^{-1},ax^{-1}-ax^{-1}) = (e,0)$$

for all $(x,a) \in G \times A$. We have shown that S is a group, and it is easy to check that

$$\left.\begin{array}{c} \lambda: A \rightarrow S \\[4pt] a \mapsto (e,a) \end{array}\right\} , \qquad \left.\begin{array}{c} \mu : S \rightarrow G \\[4pt] (x,a) \mapsto x \end{array}\right\}$$

are a monomorphism and an epimorphism, respectively, such that

$$\text{Im } \lambda = \{(e,a) \mid a \in A\} = \text{Ker } \mu \quad .$$

To verify part (iii) of Definition 2, that is, formula (3), let $x \in G$, so that $(x,0) \in \mu^{-1}(x)$, and $a \in A$, so that $a\lambda = (e,a)$. Then

$$\begin{aligned} (x,0)^{-1}(e,a)(x,0) &= (x^{-1},0)(ex,ax+0) \\[4pt] &= (x^{-1}x, 0x+ax) \\[4pt] &= (e,ax) \\[4pt] &= (ax)\lambda \quad , \end{aligned}$$

as required. The resulting extension of G by A is called their *semi-direct product*. The special case when the G-action on A is trivial (that is, $ax = a$ for all $a \in A$, $x \in G$) is none other than the direct product of G and A. One final point to notice is that the mapping

$$\left.\begin{array}{c} \sigma: G \rightarrow S \\[4pt] x \mapsto (x,0) \end{array}\right\}$$

is a homomorphism with the property $\nu\sigma = 1_G$. Because of this, we sometimes refer (loosely) to S as the split extension of G by A, in accordance with the following definition.

Definition 3 (cf. Exercise 7.3). An extension

$$K: \qquad 0 \xrightarrow{} A \xrightarrow{\iota} K \xrightarrow{\nu} G \rightarrow 1$$

of G by A is said to *split* if there is a homomorphism $\sigma\colon G \to K$
such that $\sigma\nu = 1_G$; σ is called a *splitting* for K .

Remark 4.   The extension K splits if and only if its kernel
Ker ν has a complement (see Exercise 1.5) in K . The dual condition on K , viz. the existence of a homomorphism $\rho\colon K \to A$
with $\iota\rho = 1_A$, is stronger and is equivalent to the existence of
a *normal* complement for Ker ν in K .

Remark 5.   Suppose that K ((4) above) is an extension of G
by A with splitting σ , and consider the mapping

$$\left.\begin{array}{l} \theta\colon S \to K \\[4pt] \quad (x,a) \to (x\sigma)(a\iota) \end{array}\right\} ,$$

where S is the semi-direct product of G and A . We claim that
θ is a homomorphism, noting that because of (3),

$$\nu\sigma = 1_G \Rightarrow y\sigma \in \nu^{-1}(y) \text{ , for all } y \in G$$
$$\Rightarrow (y\sigma)^{-1}(a\iota)(y\sigma) = (ay)\iota \text{ , for all } y \in G, a \in A.$$

Thus, given any $(x,a),(y,b) \in S$, we have

$$\begin{aligned}
(x,a)\theta\ (y,b)\theta &= (x\sigma)(a\iota).(y\sigma)(b\iota) \\
&= (x\sigma)(y\sigma).(y\sigma)^{-1}(a\iota)(y\sigma).(b\iota) \\
&= (x\sigma)(y\sigma)(ay)\iota(b\iota) \\
&= (xy)\sigma(ay+b)\iota \\
&= (xy,ay+b)\theta \\
&= ((x,a)(y,b))\theta \quad .
\end{aligned}$$

Furthermore,

$$a\lambda\theta = (e,a)\theta = (e\sigma)(a\iota) = a\iota$$

for all $a \in A$, and

$$(x,a)\theta\nu = ((x\sigma)(a\iota))\nu = (x\sigma\nu)(a\iota\nu) = x1_Ge = x = (x,a)\mu$$

for all $(x,a) \in S$. Thus, $\lambda\theta = \iota$, $\theta\nu = \mu$, and we have proved that the two extensions are equivalent in the following sense.

Definition 4. Two extensions K,L of G by A are called *equivalent* if there is a homomorphism $\theta : K \to L$ such that the diagram

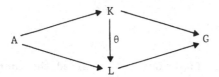

commutes.

Remark 6. That such a θ must be an *isomorphism* is a special case of the famous Five-lemma, and its proof is a nice exercise in diagram-chasing which we recommend as an exercise. We see that equivalent extensions yield isomorphic groups, though the converse is false in general (see Exercise 7).

Remark 7. Equivalence of extensions is (predictably) an equivalence relation, and the proof of this comprises another worthwhile exercise. In view of this it follows from Remark 5 that all split extensions are equivalent, and it is a simple matter to show that they in fact form an equivalence class. In so far as our overall aim in this section is the classification of extensions up to equivalence, the stockpile accumulated to date thus consists of exactly one example; we now change gear in order to get the rest.

Though an arbitrary extension

$$K: \quad 0 \to A \overset{\iota}{\to} K \overset{\nu}{\to} G \to 1$$

will not split in general, we can mimic the splitting process in the following way. By the Axiom of Choice we can choose an element $x\tau \in \nu^{-1}(x)$ for each $x \in G$, where for convenience we take

$e\tau = e \in \nu^{-1}(e)$. There results a *mapping* $\tau: G \to K$ such that $\tau\nu = 1_G$; we call τ a *transversal* for K (since Im τ is a transversal for Im ι in K in the usual sense) and denote by K_τ the extension K together with its transversal τ .

Now for any $x,y \in G$, ν sends both $x\tau y\tau$ and $(xy)\tau$ to $xy \in G$, and so these elements belong to the same coset of Ker $\nu = $ Im ι in K . It follows that $x,y \in G$ determine an element $(x,y)f \in A$ such that

$$(x\tau)(y\tau) = (xy)\tau(x,y)f\iota \quad .$$

The resulting mapping $f: G \times G \to A$ is called the *factor set* of K_τ . Note that, since $e\tau = e$, we have

$$(e,x)f = (x,e)f = 0 \tag{5}$$

for all x in G . The associative law in K and G leads to a more complicated identity. For $x,y,z \in G$,

$$(x\tau)((y\tau)(z\tau)) = x\tau \ (yz)\tau \ (y,z)f\iota$$
$$= (x(yz))\tau \ (x,yz)f\iota \ (y,z)f\iota \quad ,$$

and

$$((x\tau)(y\tau))(z\tau) = (xy)\tau \ (x,y)f\iota \ (z\tau)$$
$$= (xy)\tau \ (z\tau) \ ((x,y)f\iota)^{z\tau}$$
$$= ((xy)z)\tau \ (xy,z)f\iota \ ((x,y)fz)\iota \quad ,$$

using (3). Since ι is a momomorphism, we have the identity

$$(x,yz)f + (y,z)f = (xy,z)f + (x,y)fz \quad , \tag{6}$$

for all $x,y,z \in G$.

Definition 5. A mapping $f: G \times G \to A$ satisfying conditions (5) and (6) is called a *normalized 2-cocycle*; the set of these is

154

denoted by $Z^2(G,A)$. Pausing to observe that $Z^2(G,A)$ forms an abelian group under pointwise addition (see Exercise 8), we pass on to investigating the effect on f of changing τ . Thus, let τ, τ' be transversals (with $e\tau = e\tau' = e$) for K , with factor sets $f . f'$ respectively. Since $x\tau$ and $x\tau'$ belong to the same coset of $\text{Im } \iota$ in K , each $x \in G$ determines an $xd \in A$ by the rule

$$x\tau' = (x\tau)(xd\iota) \quad .$$

There results a mapping $d: G \to A$ such that

$$ed = 0 \quad . \tag{7}$$

We now compute (using (3)) that for all $x,y \in G$,

$$
\begin{aligned}
(xy)\tau \ (xy)d\iota \ (x,y)f'\iota &= (xy)\tau' \ (x,y)f'\iota \\
&= (x\tau')(y\tau') \\
&= x\tau \ xd\iota \ y\tau \ yd\iota \\
&= x\tau \ y\tau \ (xd\iota)^{y\tau} \ yd\iota \\
&= (xy)\tau \ (x,y)f\iota \ (xdy)\iota \ yd\iota \quad ,
\end{aligned}
$$

and since ι is a monomorphism, we have

$$(x,y)(f'-f) = yd - (xy)d + (xd)y \quad , \tag{8}$$

for all $x,y \in G$.

Definition 6.   A normalized 2-cocycle is called a *normalized 2-coboundary* if its value on $(x,y) \in G \times G$ equals the right-hand side of (8) for some d satisfying (7); the set of these is denoted by $B^2(G,A)$. Two cocycles are called *cohomologous* if they they differ by a coboundary.

Remark 8.   It is easy to see that $B^2(G,A)$ forms a subgroup of

$Z^2(G,A)$ (see Exercise 8), and we have shown that any extension of G by A determines a member of the factor group $Z^2(G,A)/B^2(G,A)$, independently of the choice of transversal. Now let K_1 and K_2 be equivalent extensions, so that we have a commutative diagram

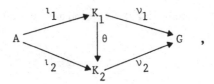

where θ is an isomorphism. Now let $\tau_1 : G \to K_1$ be a transversal for K_1, so that $\tau_2 = \tau_1\theta$ is a transversal for K_2, and let f_1, f_2 be the factor sets obtained from τ_1, τ_2 respectively. Then, for all $x, y \in G$,

$$(xy)\tau_2 \; (x,y)f_2\iota_2 = (x\tau_2) \; (y\tau_2)$$

$$= (x\tau_1\theta) \; (y\tau_1\theta)$$

$$= (x\tau_1 \; y\tau_1)\theta$$

$$= ((xy)\tau_1 \; (x,y)f_1\iota_1)\theta$$

$$= (xy)\tau_1\theta \; (x,y)f_1\iota_1\theta$$

$$= (xy)\tau_2 \; (x,y)f_1\iota_2 \quad ,$$

proving that f_1 and f_2 are equal. Letting $E(G,A)$ denote the set of equivalence classes of extensions of G by A, we thus have a well-defined mapping

$$\phi : E(G,A) \to Z^2(G,A)/B^2(G,A) \quad ,$$

induced by the formation of factor sets. We are at last in a position to prove the first (and only) theorem of this section; it entitles us to refer to $Z^2(G,A)/B^2(G,A)$ as the *group of extensions* of G by A.

Theorem 1.   *The mapping*

$$\phi: E(G,A) \to Z^2(G,A)/B^2(G,A)$$

induced by the formation of factor sets is a bijection.

Proof.   The proof that ϕ is onto, which we tackle first, is analogous to the construction performed in Example 1. Suppose that $f: G \times G \to A$ is a normalized 2-cocycle, and define a binary operation in the Cartesian product $C = G \times A$ as follows:

$$(x,a)(y,b) = (xy, ay+b + (x,y)f)$$

for all $(x,a),(y,b) \in C$. To see that this makes C into a group, note that the associative law is a consequence of (6), $(e,0)$ is right identity because of (5), and $(x^{-1}, -ax^{-1} - (x,x^{-1})f)$ is clearly a right inverse for (x,a) . It follows that this is also a left inverse for (x,a) , and we compute that

$$(x^{-1},x)f = (x,x^{-1})fx \qquad\qquad (9)$$

for all $x \in G$ (this is just (6) with $z = x, y = x^{-1}$) . As in Example 1, we define

$$\left.\begin{array}{c} \lambda: A \to C \\[4pt] a \mapsto (e,a) \end{array}\right\} \text{,} \qquad \left.\begin{array}{c} \mu: C \to G \\[4pt] (x,a) \mapsto x \end{array}\right\} \text{,}$$

and we have an extension C of G by A , provided (3) holds. To check this, let $x \in G, a \in A$ and observe that

$$(x,0)^{-1}(e,a)(x,0) = (x^{-1}, -(x,x^{-1})f)(x,ax)$$

$$= (e, -(x,x^{-1})fx + ax + (x^{-1},x)f)$$

$$= (ax)\lambda \quad ,$$

using (9). To compute a factor set for C , let $x\tau = (x,0)$ for

157

x ∈ G   and note that

$$(x\tau)(y\tau) = (x,0)(y,0)$$
$$= (xy, (x,y)f)$$
$$= (xy, 0)(e, (x,y)f)$$
$$= (xy)\tau (x,y)f\lambda \quad .$$

It follows that C has factor set f , proving that φ is onto.

   The rest of the proof consists of showing that extensions with cohomologous factor sets are equivalent. Let K_1, K_2 be extensions with transversals τ_1, τ_2 and corresponding factor sets f_1, f_2 , and assume that there is a d: G → A with ed = 0 such that

$$(x,y)(f_1-f_2) = yd - (xy)d + (xd)y \tag{10}$$

(cf. (8)) for all x,y ∈ G . We must construct a homomorphism θ such that the diagram

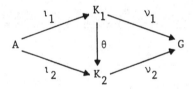

commutes. Noting that a typical element of K_1 has the form $(x\tau_1)(a\iota_1)$ for x ∈ G, a ∈ A , we define θ by

$$\left. \begin{array}{l} \theta: K_1 \to K_2 \\ (x\tau_1)(a\iota_1) \mapsto (x\tau_2)(a+xd)\iota_2 \end{array} \right\} .$$

This plainly makes the diagram commute and we merely have to prove it is a homomorphism. For all x,y ∈ G, a,b ∈ A we have

$$((x\tau_1)(a\imath_1)(y\tau_1)(b\imath_1))\theta = ((x\tau_1)(y\tau_1)\ (\imath_1)^{y\tau_1}\ (b\imath_1))\theta$$

$$= ((xy)\tau_1\ (x,y)f_1{}^{\imath_1}\ ay\imath_1\ b\imath_1)\theta$$

$$= (xy)\tau_2\ ((x,y)f_1 + ay + b + (xy)d)\imath_2\ ,$$

while on the other hand,

$$((x\tau_1)(a\imath_1))\theta\ ((y\tau_1)(b\imath_1))\theta = (x\tau_2)(a+xd)\imath_2\ (y\tau_2)(b+yd)\imath_2$$

$$= (x\tau_2)(y\tau_2)((a+xd)\imath_2)^{y\tau_2}\ (b+yd)\imath_2$$

$$= (xy)\tau_2((x,y)f_2+ay+xdy+b+yd)\imath_2\ ,$$

and these are equal by (10). Hence, θ is a homomorphism and K_1,K_2 are equivalent. This completes the proof of the theorem.

EXERCISE 1. Given a group G , a G-*set* is a set S together with an action

$$\left.\begin{array}{l} S \times G \to S \\ (s,x) \mapsto sx \end{array}\right\}$$

such that $(sx)y = s(xy)$ and $se = s$, for all $s \in S, x,y \in G$. If A is free abelian on a G-set $S \subseteq A$, prove that there is a unique G-action on A which

(i)   extends the G-action on S , and

(ii) makes A into a G-module.

EXERCISE 2. A G-set S is called *regular* if $sx \neq s$ for all $s \in S$, $x \in G\setminus\{e\}$. Invoke the Axiom of Choice to construct a subset $T \subseteq S$ such that $T \cap sG$ is a singleton for all $s \in S$. Prove that any mapping θ from T to a G-module A admits a unique extension $\theta': S \to A$ such that $(sx)\theta' = (s\theta')x$, for all $s \in S, x \in G$.

EXERCISE 3.   Given groups G and A and an action of G on A , convince yourself that a semi-direct product can be constructed as in Example 3, even when A is not abelian.

EXERCISE 4   (see Remark 4).   Prove that an extension

$$K: \quad 0 \to A \overset{\iota}{\to} K \overset{\nu}{\to} G \to 1$$

splits if and only if Ker ν has a complement in K . Show also that there is a homomorphism $\rho: K \to A$ such that $\iota\rho = 1_A$ if and only if Ker ν has a normal complement in K and deduce that in this case, K is isomorphic to the direct product of G and A .

EXERCISE 5   (see Remark 6).   Prove that if K,L are equivalent extensions of G by A , then K,L are isomorphic groups.

EXERCISE 6.   Prove that equivalence of extensions is an equivalence relation.

EXERCISE 7.   If G is a group and A is a finite G-module, prove that every element of $Z^2(G,A)/B^2(G,A)$ has order dividing |A| . By taking $G = A = Z_3$ with trivial action, prove the existence of inequivalent extensions K,L with isomorphic groups K,L .

EXERCISE 8.   Given a group G and a G-module A , prove that $Z^2(G,A)$ is a group under pointwise addition, and that $B^2(G,A)$ is a subgroup of it.

EXERCISE 9   (cf. Exercise 1.5).   Let F be a free group and A an F-module. Prove that $B^2(F,A) = Z^2(F,A)$.

EXERCISE 10   (cf. Exercises 1.8 and 12.1).   Let S be the split extension of $G = <c \mid c^2>$ by $A = <a,b \mid >$, where the action of c is induced by transposing a and b . Prove that S is just the group

$$B = <a,b,c \mid a^c = b, \; b^c = a, \; c^2 = e> \quad .$$

EXERCISE 11 (cf. Exercise 7.7). Let p be an odd prime, let $a,b \in N$, and consider the action

$$
\left.
\begin{array}{l}
<y \mid y^{p^b}> \;\to\; \mathrm{Aut} <x \mid x^{p^{a+b}}> \\[2mm]
\phantom{<y \mid y} y \;\mapsto\; (x \mapsto x^{1+p^a})
\end{array}
\right\} \quad .
$$

Prove that the resulting split extension is just the metacyclic group

$$G = <x,y \mid x^{p^{a+b}} = e, \; y^{-1}xy = x^{1+p^a}, \; y^{p^b} = e> \quad .$$

§18. Teach yourself cohomology of groups

We aim to put the ideas of the previous section in a broader context, and in particular to explain how the group $Z^2(G,A)/B^2(G,A)$ of extensions of a group G by a G-module A arises in a natural way. To do this requires a more detailed study of abelian groups, or of modules over an arbitrary ring. We devote the first part of this section to basic definitions and elementary results, whose proofs (see Exercises 1-5) the reader is invited to work out as he goes along. The multiplicative group G is fixed throughout, and G-modules are written additively with G acting on the right as in Definition 17.1.

Given a G-module B, a G-*submodule* of B is a subset A of B such that

(i) A is a subgroup of B, and

(ii) A is closed under the G-action,

$$AG = \{ax \mid a \in A, \; x \in G\} \subseteq A \quad .$$

Because of (i), we can form the factor group B/A, and the definition

$$(A+b)x = A + bx , \quad b \in B, \; x \in G ,$$

makes B/A into a G-module called the *factor module* of B by A
(see Exercise 1).

Given G-modules A and Y , a mapping $\theta: A \to Y$ is called a
G-*homomorphism* if

(i) θ is a homomorphism of groups, $(a+a')\theta = a\theta + a'\theta$ $(a,a' \in A)$ and

(ii) θ commutes with the G-action, $(ax)\theta = (a\theta)x$ $(a \in A, x \in G)$.

The *image* and *kernel* of a G-homomorphism $\theta: A \to Y$, given by:

$$\text{Im } \theta = \{a\theta \mid a \in A\} \ , \quad \text{Ker } \theta = \{a \in A \mid a\theta = 0\} \ ,$$

are G-submodules of Y,A respectively (see Exercises 3,4).

We write $A^* = \text{Hom}_G(A,Y)$ for the set of G-homomorphisms from
A to Y . Under pointwise addition

$$\left. \begin{aligned} (\theta + \phi): A &\to Y \\ a &\mapsto a\theta + a\phi \end{aligned} \right\} ,$$

$\text{Hom}_G(A,Y)$ forms an abelian group (see Exercise 5). Given a G-
module Y and a G-homomorphism $\theta: A \to B$, we define
$\theta^* = \text{Hom}_G(\theta,Y)$ by

$$\left. \begin{aligned} \theta^*: \text{Hom}_G(B,Y) &\to \text{Hom}_G(A,Y) \\ \alpha &\mapsto \theta\alpha \end{aligned} \right\} ,$$

and note that this is a homomorphism of abelian groups (Exercise 6).
The operation * preserves sums, composites and identity mappings.
These are called *functorial properties*, expressed by

$$(\theta + \phi)^* = \theta^* + \phi^* , \quad (\phi\psi)^* = \psi^*\phi^* , \quad (1_A)^* = 1_{A^*} ,$$

where θ, ϕ, ψ are G-homomorphisms with appropriate domains and
codomains.

An *exact sequence* of G-modules is a collection of
G-modules $\{A_n \mid n \in Z\}$ and G-homomorphisms $\{\theta_n: A_n \to A_{n-1} \mid n \in Z\}$
such that $\text{Im } \theta_n = \text{Ker } \theta_{n-1}$ for all $n \in Z$. Note that

Im $\theta_n \subseteq$ Ker θ_{n-1} if and only if $\theta_n\theta_{n-1} = 0$. A *short exact sequence* (cf. Definition 17.2) is an exact sequence in which all but three consecutive terms are zero. A typical short exact sequence is usually written in the form

$$0 \to A \overset{\theta}{\to} B \overset{\phi}{\to} C \to 0 \ , \tag{1}$$

where the exactness means that:

θ is one-to-one, $\theta\phi = 0$, Im $\theta \supseteq$ Ker ϕ , ϕ is onto.

For example, if A is a G-submodule of B , we have a short exact sequence

$$0 \to A \overset{inc}{\to} B \overset{nat}{\to} B/A \to 0 \ ,$$

and if A,B are any G-modules, we have a short exact sequence

$$\left. \begin{array}{c} 0 \to A \to A \oplus B \to B \to 0 \\ a \mapsto (a,0) \\ (a,b) \mapsto b \end{array} \right\} \ .$$

The latter is an example of a split short exact sequence in the following sense: a typical short exact sequence (1) is said to *split* (cf. Definition 17.3) if there is a G-homomorphism $\sigma\colon C \to B$ such that $\sigma\phi = 1_C$. Two equivalent definitions of splitting are given in Exercise 8.

We now consider the effect of applying the operator $\mathrm{Hom}_G(\ ,Y)$ to the short exact sequence (1). Since ϕ^* is one-to-one and Im $\phi^* =$ Ker θ^* (see Exercise 9), we write

$$0 \to C^* \overset{\phi^*}{\to} B^* \overset{\theta^*}{\to} A^* \ , \tag{1*}$$

to emphasize the fact that θ^* may not be onto (Exercise 10). Sometimes it is however, for example when (1) splits (Exercise 11).

Definition.   A G-module F is said to be free on a subset $S \subseteq F$
if, for any G-module A and any mapping $\theta: S \to A$, there is a
unique G-homomorphism $\theta': F \to A$ extending θ .

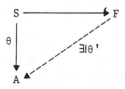

S is called a G-*basis* for F , and $|S|$ the *rank* of F .

Remark 1.   There is a close analogy between free G-modules and
free groups, and properties corresponding to those described in
§1 can be derived in similar fashion. For example, if F is
G-free on S , then no proper G submodule of F contains S ,
that is, S generates F as a G-module (see Exercise 13).

Remark 2.   Let F,A be G-modules with F free on S , and let
$\iota: S \to F$ denote inclusion. Consider the *restriction* mapping

$$\left.\begin{array}{l} \iota^*: \mathrm{Hom}_G(F,A) \;\to\; \mathrm{Map}(S,A) \\[2mm] \qquad\quad \alpha \;\mapsto\; \iota\alpha \end{array}\right\} ,$$

where $\mathrm{Map}(S,A)$ just consists of set-functions. The definition
of freeness then simply asserts that ι^* is a *bijection*; in fact,
the correspondence $\theta \to \theta'$ in the definition is the inverse
of ι^* . In this situation, we tend to identify $\mathrm{Hom}_G(F,A)$ and
$\mathrm{Map}(S,A)$.

Remark 3.   If F_i is free on S_i (i = 1,2) and $F_1 \cong F_2$ as
G-modules, it follows that $\mathrm{Map}(S_1,A)$ and $\mathrm{Map}(S_2,A)$ have the
same number of elements for any G-module A . Taking $A = Z_2$
with the trivial G-action, we deduce that $2^{|S_1|} = 2^{|S_2|}$, whence
$|S_1| = |S_2|$, and the rank of a free G-module is well defined.

Remark 4.   In similar vein, if F_i is free on S_i (i = 1,2) and

164

$|S_1| = |S_2|$, then F_1 and F_2 are isomorphic as G-modules (see Exercise 14).

Remark 5.   We proceed to question the existence of free G-modules. Given any set S , let F be the free abelian group on the Cartesian product $S \times G$.   For any $x \in G$, define

$$\left.\begin{array}{l} \alpha_x \colon S \times G \to F \\ \\ \quad (s,z) \mapsto (s,zx) \end{array}\right\} ,$$

and for any $a \in F$, we define $ax = a\alpha_x' \in F$, where α_x' is the unique homomorphism $F \to F$ extending α_x .   To verify the three module axioms, note first that this action is linear, since α_x' is a homomorphism of groups.   Next, α_e' fixes the generators $S \times G$ and thus fixes everything.   Finally, the homomorphisms $\alpha_x'\alpha_y'$, α_{xy}' map $(s,z) \in S \times G$ to

$$(s,(zx)y) , \quad (s,z(xy))$$

respectively, and thus are equal.

We now relabel $(s,e) \in S \times \{e\}$ as s , so that $(s,z) = (s,e)z = sz$ for $z \in G$, and F is the free abelian group on the set $SG = \{sz \mid s \in S, z \in G\}$.   To prove that F is G-free on S , let A be any G-module and $\theta \colon S \to A$ any mapping. We define

$$\left.\begin{array}{l} \theta' \colon SG \to A \\ \\ \quad sz \mapsto (s\theta)z \end{array}\right\} ,$$

and let $\theta'' \colon F \to A$ be the unique homomorphism (of groups) extending θ' .

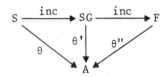

Now for any $x, z \in G$, $s \in S$, we have

$$((sz)x)\theta'' = (s(zx))\theta' = (s\theta)zx = (s\theta z)x = (sz)\theta'x \ ,$$

so that $x\theta'' = \theta''x$, since they agree on a Z-basis. Hence, θ'' is a G-homomorphism. Finally, if two G-homomorphisms agree with θ on S, they agree with each other on a set of G-generators and thus are equal.

Example 1. For n a non-negative integer, we let the Cartesian product G_n of n copies of G play the role of S in the above construction. As $x_0, \ldots, x_{n-1}$ and $x_0, \ldots, x_n$ range independently over G, the elements

$$(x_0, \ldots, x_{n-1}) \quad \text{and} \quad (x_0, \ldots, x_{n-1})x_n$$

yield a G-basis and a Z-basis respectively for the resulting G-free module F_n. Defining

$$[x_0, \ldots, x_n] = (x_0 x_n^{-1}, \ldots, x_{n-1} x_n^{-1})x_n \ , \tag{2}$$

these bases can be rewritten as

$$[x_0, \ldots, x_{n-1}, e] \quad \text{and} \quad [x_0, \ldots, x_n]$$

respectively, since as $x_0, \ldots, x_n$ range independently over G so do the $n+1$ coordinates of

$$[x_0 x_n, \ldots, x_{n-1} x_n, x_n] = (x_0, \ldots, x_{n-1})x_n \ .$$

In the new notation, the G-action on the Z-basis is given by

$$[x_0, \ldots, x_n]x = [x_0 x, \ldots, x_n x] \ . \tag{3}$$

For $n = -1$, we interpret F_{-1} as the free abelian group on the singleton $[\ ]$, with trivial G-action.

166

We now define $\partial_n\colon F_n \to F_{n-1}$ to be the unique homomorphism (of groups) extending the mapping

$$[x_o,\ldots,x_n] \mapsto \sum_{i=0}^{n} (-1)^i [x_o,\ldots,\hat{x}_i,\ldots,x_n] \tag{4}$$

on the Z-basis, where $\wedge$ denotes the omission of the coordinate beneath it. Since ∂_n clearly commutes with the G-action on the Z-basis, it must be a G-homomorphism for all $n \geq 0$. Note that since $[\ ] = [e]\partial_o \in \mathrm{Im}\ \partial_o$, and generates F_{-1}, ∂_o is onto.

Theorem 1. *With the above notation, the sequence*

$$F\colon \quad \cdots \to F_n \overset{\partial_n}{\to} F_{n-1} \to \cdots \to F_o \overset{\partial_o}{\to} F_{-1} \to 0 \tag{5}$$

is exact.

Proof. Having just observed that ∂_o is onto, we have two things to check, namely (i) $\partial_{n+1}\partial_n = 0$ and (ii) $\mathrm{Ker}\ \partial_n \subseteq \mathrm{Im}\ \partial_{n+1}$, for all $n \geq 0$. Since the ∂_n are homomorphisms, (i) will be proved if we can show that $\partial_{n+1}\partial_n$ vanishes on a typical element $[x_o,\ldots,x_{n+1}]$ of the Z-basis for F_{n+1}. Well, for all $n \geq 0$ we have:

$$[x_o,\ldots,x_{n+1}]\partial_{n+1}\partial_n$$

$$= (\sum_{i=0}^{n+1} (-1)^i [x_o,\ldots,\hat{x}_i,\ldots,x_{n+1}])\partial_n$$

$$= \sum_{i=0}^{n+1} (-1)^i [x_o,\ldots,\hat{x}_i,\ldots,x_{n+1}]\partial_n$$

$$= \sum_{i=0}^{n+1} (-1)^i (\sum_{j<i} (-1)^j [x_o,\ldots,\hat{x}_j,\ldots,\hat{x}_i,\ldots,x_{n+1}]$$

$$\qquad\qquad + \sum_{j>i} (-1)^{j-1}[x_o,\ldots,\hat{x}_i,\ldots,\hat{x}_j,\ldots,x_{n+1}])$$

$$= \sum_{0\leq j<i\leq n+1} (-1)^{i+j} [x_o,\ldots,\hat{x}_j,\ldots,\hat{x}_i,\ldots,x_{n+1}]$$

$$+ \sum_{0\leq i<j\leq n+1} (-1)^{i+j-1}[x_o,\ldots,\hat{x}_i,\ldots,\hat{x}_j,\ldots,x_{n+1}] \quad,$$

and by interchanging the dummy variables i and j in the second sum, we see that this is zero.

Oddly enough, the proof of (ii) is even simpler, provided we invoke a neat little trick. For each, $n \geq 0$, define

$$\left. \begin{array}{l} \beta_n : F_{n-1} \to F_n \\[1mm] [x_0, \ldots, x_{n-1}] \mapsto [e, x_0, \ldots, x_{n-1}] \end{array} \right\},$$

a homomorphism of abelian groups (though not of G-modules). From the picture

$$\cdots \stackrel[{\beta_{n+1}}]{\partial_{n+1}}{\rightleftarrows} F_{n+1} \stackrel[{\beta_n}]{\partial_n}{\rightleftarrows} F_n \rightleftarrows F_{n-1} \rightleftarrows \cdots \rightleftarrows F_0 \stackrel[{\beta_0}]{\partial_0}{\rightleftarrows} F_{-1} \to 0 \ ,$$

we have two additive endomorphisms, $\partial_n \beta_n$ and $\beta_{n+1} \partial_{n+1}$ of F_n , for each $n \geq 0$; we apply their sum to a typical member $[x_0, \ldots, x_n] \in G_{n+1}$:

$$[x_0, \ldots, x_n](\partial_n \beta_n + \beta_{n+1} \partial_{n+1})$$

$$= \sum_{i=0}^{n} (-1)^i [x_0, \ldots, \hat{x}_i, \ldots, x_n]\beta_n + [e, x_0, \ldots, x_n]\partial_{n+1}$$

$$= \sum_{i=0}^{n} (-1)^i [e, x_0, \ldots, \hat{x}_i, \ldots, x_n] + [x_0, \ldots, x_n]$$

$$\qquad + \sum_{i=0}^{n} (-1)^{i+1} [e, x_0, \ldots, \hat{x}_i, \ldots, x_n]$$

$$= [x_0, \ldots, x_n] \ .$$

Since $\partial_n \beta_n + \beta_{n+1} \partial_{n+1}$ fixes a basis, it must be the identity on F_n , and so for any $a \in \mathrm{Ker}\ \partial_n$, we have:

$$a = a\partial_n \beta_n + a\beta_{n+1}\partial_{n+1} = (a\beta_{n+1})\partial_{n+1} \in \mathrm{Im}\ \partial_{n+1} \ ,$$

as required.

An exact sequence of free G-modules ending in Z is called a *G-free resolution of* Z ; the example (5) just constructed is called the *bar resolution*, and was inspired by homological con-

siderations in algebraic topology. Our next aim is to take a
G-module A and apply the operator $* = \mathrm{Hom}_G(\ ,A)$ to the resolution (5), but before doing so, we adjust our notation for the
G-bases used for the F_n's . As $x_1,\ldots,x_n$ range independently
over G $(n \geq 0)$, so do the n coordinates of

$$(x_1,\ldots,x_n) := [x_1\ldots x_n, x_2\ldots x_n,\ldots,x_{n-1}x_n,x_n,e] \quad ,$$

so that these elements comprise a G-basis for F_n . Taking a
fixed G-module A , the elements of

$$F_n^* = \mathrm{Hom}_G(F_n,A)$$

correspond to set-functions from this G-basis into A in accordance with Remark 2 above. We now compute the effect of
$\partial_n^* = \mathrm{Hom}_G(\partial_n,A)$ on a typical member of $\mathrm{Hom}(G^{\times(n-1)},A)$ in the
sequence

$$F^* \qquad \ldots \ \leftarrow \ F_n^* \ \overset{\partial_n^*}{\leftarrow} \ F_{n-1}^* \ \leftarrow \ \ldots \ \leftarrow \ F_o^* \quad .$$

Lemma 1. *For* $n \geq 1$ *and* $f \in \mathrm{Hom}(G^{\times(n-1)},A)$, *the mapping*
$f\partial_n^* \in \mathrm{Hom}(G^{\times n},A)$ *is given by:*

$$(x_1,\ldots,x_n)f\partial_n^* = (x_2,\ldots,x_n)f + \sum_{i=1}^{n-1} (-1)^i(x_1,\ldots,x_ix_{i+1},\ldots,x_n)f$$

$$+ (-1)^n(x_1,\ldots,x_{n-1})fx_n \quad . \tag{6}$$

Proof. We compute that

$$(x_1,\ldots,x_n)f\partial_n^* = [x_1\ldots x_n,\ldots,x_n,e]\partial_n f$$

$$= \sum_{i=1}^{n} (-1)^{i-1}[x_1\ldots x_n,\ldots,\overset{\frown}{x_i\ldots x_n},\ldots,x_n,e]f$$

$$+ (-1)^n[x_1\ldots x_n,\ldots,x_n]f$$

$$= (x_2, \ldots, x_n)f + \sum_{i=2}^{n} (-1)^{i-1}(x_1, \ldots, x_{i-1}x_i, \ldots, x_n)f$$

$$+ (-1)^n([x_1 \ldots x_{n-1}, \ldots, e]x_n)f$$

$$= (x_2, \ldots, x_n)f + \sum_{i=1}^{n-1} (-1)^i (x_1, \ldots, x_i x_{i+1}, \ldots, x_n)f$$

$$+ (-1)^n(x_1, \ldots, x_{n-1})fx_n \quad ,$$

replacing i by $i+1$ in the summation.

By force of tradition, we define $\partial_o^* = 0$, and observe that by the functorial properties of $*$, we have

$$\partial_n^* \partial_{n+1}^* = (\partial_{n+1} \partial_n)^* = 0^* = 0 \quad ,$$

and the question arises as to by how much the sequence F^* fails to be exact. The answer is provided by the groups

$$H^n(G,A) = \operatorname{Ker} \partial_{n+1}^* / \operatorname{Im} \partial_n^* \quad , \quad n \geq 0 \quad ,$$

called the *cohomology groups of* G *with coefficients in* A. The methods of homological algebra show that these groups are independent of the choice of free G-resolution used to define them, though it is beyond our scope to prove this here. We content ourselves instead with a closer look at what happens for small values of n.

Example 0. The G-basis for F_o consists of the singleton $[e] = ( )$, so that mappings from this into A simply correspond to members of A:

$$\left. \begin{array}{c} F_o^* \to A \\ f \mapsto ( )f \end{array} \right\} .$$

With this identification, a member $a = ( )f \in A$ lies in $\operatorname{Ker} \partial_1^* = H^o(G,A)$ if and only if, for all $x \in G$,

$$0 = (x)f\partial_1^* = (\ )f - (\ )fx = a - ax \quad,$$

using Lemma 1. This simply asserts that a belongs to the G-trivial submodule A^G of elements of A fixed by all $x \in G$. In particular $H^0(G,Z) = Z^G = Z$.

Example 1. A mapping $f: G \to A$ is in Ker ∂_2^* if and only if, for all $x,y \in G$:

$$0 = (x,y)(f\partial_2^*) = (y)f - (xy)f + (x)fy \quad,$$

that is (omitting the brackets),

$$(xy)f = yf + (xf)y \quad.$$

Such mappings are called *crossed homomorphisms* from G to A , and are just ordinary homomorphisms when the G-action is trivial. Similarly, $f: G \to A$ is in Im ∂_1^* if and only if there is an $a = (\ )f' \in A$ such that, for all $x \in G$,

$$xf = (x)(f'\partial_1^*) = (\ )f' - (\ )f'x = a(e-x) \quad.$$

Crossed homomorphisms of this type are called *principal*. When the G-action on A is trivial, Im $\partial_1^* = \{0\}$ and $H^1(G,A)$ just consists of group homomorphisms. As a special case, $H^1(G,Z)$ is trivial for finite G .

Example 2. A mapping $f: G \times G \to A$ is in Ker ∂_3^* if and only if, for all $x,y,z \in G$:

$$0 = (x,y,z)(f\partial_3^*) = (y,z)f - (xy,z)f + (x,yz)f - (x,y)fz \quad, \qquad (7)$$

which is precisely the condition 17.(6) that f be a 2-cocycle. Similarly, $f \in$ Im ∂_2^* if and only if there is a mapping $d: G \to A$ such that, for all $x,y \in G$,

$$(x,y)f = (x,y)(d\partial_2^*) = (y)d - (xy)d + (x)dy \quad , \tag{8}$$

which is the condition 17.(8) that f be a 2-coboundary. Thus, $Z^2(G,A)$ consists of the *normalized* members of $\text{Ker } \partial_3^*$, and $B^2(G,A)$ consists of the members of $\text{Im } \partial_2^*$ that are images of *normalized* mappings: $G \to A$. We now complete the final step in the identification of $H^2(G,A)$ with the group $E(G,A)\phi$ of extensions of G by A.

Theorem 2. *The groups* $Z^2(G,A)/B^2(G,A)$ *and* $H^2(G,A)$ *are isomorphic.*

Proof. The inclusion of $Z^2(G,A)$ in $\text{Ker } \partial_3^*$ obviously maps $B^2(G,A)$ into $\text{Im } \partial_2^*$, and we merely have to show that the resulting induced homomorphism

$$\psi: \frac{Z^2(G,A)}{B^2(G,A)} \to \frac{\text{Ker } \partial_3^*}{\text{Im } \partial_2^*}$$

is a bijection. Now

$$\text{Ker } \psi = \frac{Z^2(G,A) \cap \text{Im } \partial_2^*}{B^2(G,A)} \quad , \quad \text{Im } \psi = \frac{Z^2(G,A) + \text{Im } \partial_2^*}{\text{Im } \partial_2^*} \quad ,$$

so we have to show that

(i) $Z^2(G,A) \cap \text{Im } \partial_2^* \subseteq B^2(G,A)$, and

(ii) $Z^2(G,A) + \text{Im } \partial_2^* \supseteq \text{Ker } \partial_3^*$.

To prove (i), let $f: G \times G \to A$ be a normalized mapping satisfying (7) and (8). Then, from (8),

$$0 = (e,e)f = (e)d - (ee)d + (e)de = (e)d \quad ,$$

showing that $f = d\partial_2^*$ with $d: G \to A$ normalized, that is, $f \in B^2(G,A)$. For (ii), let $f: G \times G \to A$ satisfy (7), so that, taking $(x,y,z) = (e,e,y),(x,e,e)$ respectively, we have

172

$$0 = (e,y)f - (e,e)fy = (e,e)f - (x,e)f \quad . \tag{9}$$

Now let $d: G \to A$ be the constant mapping sending all x to $(e,e)f$. The mapping $f' = f - d\partial_2^*: G \times G \to A$ obviously belongs to $\operatorname{Ker} \partial_3^*$, and is given by:

$$(x,y)f' = (x,y)f - (e,e)fy \quad .$$

Because of (9), f' is normalized, and we have

$$f = f' + d\partial_2^* \in Z^2(G,A) + \operatorname{Im} \partial_2^* \quad ,$$

as required.

Example 3. When G is finite and Z has the trivial G-action, the group $H^3(G,Z)$ also boasts a purely group-theoretical significance. Given a finite presentation $<X \mid R>$ for G , let F be the free group on X , and $\bar{R}$ the normal closure of R in F . Denoting by $[F,\bar{R}]$ the subgroup of F generated by the commutators $\{[f,r] \mid f \in F, r \in \bar{R}\}$, it is clear that $[F,\bar{R}] \subseteq F'$ and also $[F,\bar{R}] \subseteq \bar{R}$, since $\bar{R}$ is normal. Since $\bar{R}' \subseteq [F,\bar{R}]$, the group

$$M(G) = (F' \cap \bar{R})/[F,\bar{R}]$$

is abelian; it turns out that $M(G)$ is finite, and is independent of the presentation used to define it. $M(G)$ thus depends only on G , and is called the *Schur multiplicator* of G (cf. the beginning of Chapter III). Though it is beyond our scope to prove it here, the groups $M(G)$ and $H^3(G,Z)$ are isomorphic for finite G .

EXERCISE 1. Given an inclusion of G-modules $A \subseteq B$, check that the definition of the factor module B/A makes sense.

EXERCISE 2. Concoct a reasonable definition of the direct sum of

G-modules $A_1, \ldots, A_n$.

EXERCISE 3. Given a G-homomorphism $\theta: A \to Y$, prove that
Im θ, Ker θ are submodules of Y,A , respectively.

EXERCISE 4. Think of some isomorphism theorems for abelian
groups, then state and prove their analogues for G-modules.

EXERCISE 5. Check that $\mathrm{Hom}_G(A,Y)$ forms an abelian group under
pointwise addition.

EXERCISE 6. Given G-homomorphisms $\theta: A \to B$, $\alpha, \beta: B \to Y$, check
that $\alpha\theta^* := \theta\alpha \in \mathrm{Hom}_G(A,Y)$, and that $(\alpha+\beta)\theta^* = \alpha\theta^* + \beta\theta^*$.

EXERCISE 7. Check the three functorial properties of
$* = \mathrm{Hom}_G(\ ,Y)$.

EXERCISE 8. Prove that the following three conditions on the
short exact sequence (1) are equivalent:
a)   there is a $\sigma \in \mathrm{Hom}_G(C,B)$ such that $\sigma\phi = 1_C$,
b)   Im θ = Ker ϕ is complemented as a G-submodule of B ,
c)   there is a $\rho \in \mathrm{Hom}_G(B,A)$ such that $\theta\rho = 1_A$.

EXERCISE 9. Prove that the result of applying the operator
$* = \mathrm{Hom}_G(\ ,Y)$ to the short exact sequence (1) is a sequence (1)*
such that

$$0 = \mathrm{Ker}\ \phi^* , \quad \mathrm{Im}\ \phi^* = \mathrm{Ker}\ \theta^* .$$

EXERCISE 10. Noting that E-modules are nothing but abelian
groups, concoct an example with $A = C = Y = Z_2$ to show that θ^*
may not be onto in (1)*.

EXERCISE 11. Given that the short exact sequence (1) is split
by $\sigma: C \to B$, prove that θ^* is onto in (1)*, and use Exercise 8
to show that the resulting short exact sequence also splits.

174

EXERCISE 12 (cf. Exercise 1.1).   If   S   is a subset of a G-
module   A , prove that   <S>   consists of finite sums of members
of   SG .

EXERCISE 13 (cf. Theorem 1.1(i)).   If   F   is a G-module free
on   S , prove that   F   is generated by   S .

EXERCISE 14.   If F_1, F_2 are G-modules with F_1 free, then
$F_1 \cong F_2$ if and only if F_2 is free and has the same rank as F_1 .

EXERCISE 15.   Prove that a free G-module of rank $n \in N$ is isomorphic to the direct sum of   n   copies of a free G-module of
rank   1.

EXERCISE 16.   Let   A,B   be G-modules and $\theta: A \to B$ an additive
mapping. If   S   generates   A   as an abelian group and
$(sx)\theta = (s\theta)x$ for all $s \in S$, $x \in G$, prove that θ is a G-
homomorphism. Is it enough to assume that   S   generates   A   as
a G-module?

§19. Local cohomology and p-groups

   The construction of the bar resolution carried out in the
previous section can be modified as follows. Whereas F_n was
defined to be the group of all finite Z-linear combinations of
members of $(G^{\times n})G$, we now define ${}^p F_n$ to be the set of all
finite k-linear combinations of $(G^{\times n})G$, where $k = Z/pZ$ is the
field of   p   elements (p   a fixed prime). ${}^p F_n$ is thus a vector
space over   k   having the elements of $(G^{\times n})G$ as a basis in the
usual sense of linear algebra (since   k   is a field). Now pF_n
is clearly a G-submodule of F_n , and we can think of ${}^p F_n$ as
the factor module F_n/pF_n . Proceeding exactly as before, we
rename the k-basis as in 18.(2) with G-action as in 18.(3), and
define ${}^p \partial_n$ the extension by k-linearity of 18.(4). That the
resulting sequence ${}^p F$ (see 18.(5)) is exact is proved in
exactly the same way as Theorem 18.1.

Though the G-modules ^PF_n are no longer free (for $n \geq 0$), they are still 'relatively free' (G-free mod p) in the following sense. Firstly, p annihilates ^PF_n . Secondly, given a G-module A and a mapping $\theta: (G^{\times n}) \to A$, consider the following diagram:

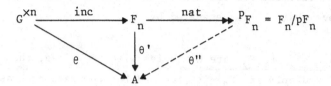

where θ' is the unique G-homomorphism extending θ . Now there will be a θ'' making this commute provided that $pF_n \subseteq \mathrm{Ker}\ \theta'$, and since $(pF_n)\theta' \subseteq pA$, this is guaranteed if we assume that $pA = 0$, that is, A is a vector space over k . When this happens, the induced G-homomorphism $\theta'': {}^PF_n \to A$ is unique (since nat is onto). Thus, any $\theta: G^{\times n} \to A$ admits a unique extension to ^PF_n *provided that* p *annihilates* A .

Assuming that $pA = 0$, we can thus identify $^PF_n^*$ ($n \geq 0$) with $\mathrm{Map}(G^{\times n}, A)$ exactly as before, and the formula for $^P\partial_n^*$ is identical to that given in 18.(6) for ∂_n^* . It follows that when $pA = 0$, $H^n(G,A)$ is isomorphic to $\mathrm{Ker}\ {}^P\partial_{n+1}^*\ /\ \mathrm{Im}\ {}^P\partial_n^*$. This approach has many advantages, for example, when G is finite, so are all the ^PF_n . If A is also finite, then so are all the $^PF_n^*$, whence the $H^n(G,A)$ are finite too.

We proceed to examine the whole picture from a slightly different point of view. Let kG denote the vector space over k with basis G (cf. Exercise 18.12), so that a typical element of kG has the form

$$a = \sum_{x \in G}{}' a_x x \ ,$$

where the $a_x \in k$ and the $'$ denotes that only finitely many of them are non-zero. Now the definition $ay = \sum_{x \in G}{}' a_x (xy)$ clearly makes kG into a G-module, and this can be extended to an action of kG by k-linearity, that is, if

$$\alpha = \sum_{x \in G}' a_x x , \qquad \beta = \sum_{y \in G}' b_y y ,$$

we define

$$\alpha\beta = \sum_{y \in G}' b_y (\sum_{x \in G}' a_x (xy)) = \sum_{z \in G}' c_z z ,$$

where

$$c_z = \sum_{xy=z}' b_y a_x = \sum_{y=x^{-1}z}' a_x b_y = \sum_{x \in G}' a_x b_{x^{-1}z} . \tag{1}$$

It is easy to verify (Exercise 1) that this definition imbues kG with the structure of a ring with identity $1e = 1$.

Definition. We call kG the *group ring of* G *over* k .

Remark 1. If A is a G-module with $pA = 0$, we let $\alpha = \Sigma' a_x x \in kG$ operate on $a \in A$ by the rule:

$$a\alpha = \sum_x' a_x (ax) .$$

One readily checks (Exercise 2), that for all $a,b \in A$, $\alpha,\beta \in kG$,

$$(a+b)\alpha = a\alpha + b\alpha, \ a(\alpha+\beta) = a\alpha + a\beta, \ a(\alpha\beta) = (a\alpha)\beta, \ a1 = a,$$

so that A becomes a (right, unital) kG-module in the usual ring-theoretic sense. Conversely, any kG-module becomes a G-module simply by restricting the operators to the basis $G \subseteq kG$.

Remark 2. Just as with any ring, the internal multiplication makes kG into a module over itself. The result is nothing other than $^p F_o$, since the mapping

$$\left. \begin{array}{l} kG \rightarrow {}^p F_o \\ \sum_x' a_x x \mapsto \sum_x' a_x [\ ]x \end{array} \right\}$$

is easily seen to be a bijective G-homomorphism (Exercise 3).
Our notion of 'G-free mod p' thus coincides with 'kG-free' in the
ring-theoretic sense, provided the rank is 1. That this is
true for arbitrary ranks is an easy exercise (Exercise 4). It
follows that a G-module which is free mod p is just a direct
sum of copies of kG.

Remark 3.   We have shown above that when A is a G-module with
pA = 0 (that is, a kG-module), the groups $H^n(G,A)$ can be
computed as kernels/images in the sequence obtained by applying
the operator $\text{Hom}_G(\ ,A)$ to a kG-free resolution of k. It
turns out that the $H^n(G,A)$ are independent of the resolution
used (cf. the remark following their definition on p.170); we
shall later make crucial use of this fact.

Remark 4.   We now turn our attention to the G-homomorphism
$^P\partial_o : {}^P F_o \to {}^P F_{-1}$. Regarding ${}^P F_o, {}^P F_{-1}$ as kG,k respectively,
we refer to $^P\partial_o$ as ε , given by

$$\left. \begin{array}{l} \varepsilon: kG \to k \\[6pt] \sum{}' a_x x \mapsto \sum{}' a_x \end{array} \right\} .$$

It is easy to show (Exercise 5) that ε is a homomorphism of
rings and of kG-modules, so that $U = \text{Ker } \varepsilon$ is an ideal of kG.
ε is called the *augmentation* mapping, and U the *augmentation
ideal of* kG. U is variously referred to as the Magnus ideal,
difference ideal or fundamental ideal of kG. The set
$\{(e-x) \mid x \in G \backslash \{e\}\}$ comprises a k-basis for U (Exercise 6), so
that $\dim_k U = |G| - 1$, and U is a maximal right ideal of kG
with $kG/U \cong k$. It follows that k is an *irreducible* kG-module
in the ring-theoretic sense, that is, it has exactly two kG-sub-
modules (itself and O); for, as its k-dimension is 1, it has
no proper non-trivial k-subspace. The following striking result
is the key to the remainder of this section.

Theorem 1. *When* G *is a finite* p-*group and* k *is the field of* p *elements,* k *is the only irreducible* kG-*module.*

Proof (J.-P. Serre). Let A be an irreducible kG-module and let $0 \neq a \in A$. The set $\{ax \mid x \in G\}$ spans a subspace B of A with $m = \dim_k B \le |G|$. It is clear that B is closed under post-multiplication by the elements of G (cf. Exercise 18.12), so that the action of G permutes the members of B . If $\ell_1, \ldots, \ell_n$ are the lengths of the orbits under this action, we have

$$p^m = \ell_1 + \ldots + \ell_n \ ,$$

and each ℓ_i is a p-power (see Exercise 7). Since one of the orbits consists of 0 alone, there is at least one other singleton orbit, $\{b\}$ say. Thus, $bx = b$ for all $x \in G$ and $b \neq 0$, whence $C = \{\lambda b \mid \lambda \in k\}$ comprises a kG-submodule of A . Since $\dim_k C = 1$ and G acts trivially on C , $C \cong k$, and since $C \neq 0$ and A is irreducible, $C = A$. Hence, $A \cong k$ as required.

Remark 5. If A is a kG-module (for arbitrary G), we let AU denote the k-subspace of A spanned by the set $\{au \mid a \in A, u \in U\}$, and observe that AU is a kG-submodule of A . Because of the structure of U (Exercise 6), AU is actually spanned by elements of the form $a(e-x)$, $a \in A$, $x \in G$, so that when G acts trivially on A , we have $AU = 0$. Conversely, if $AU = 0$, then $a(e-x) = 0$ for all $a \in A$, $x \in G$, and the G-action is trivial. It follows that G acts trivially on A/AU , and that whenever B is a submodule of A such that the G-action on A/B is trivial, we must have $AU \subseteq B$. Thus we refer to A/AU as the largest G-trivial factor module of A , and sometimes denote it by A_G (cf. Example 18.0).

Remark 6. In view of Theorem 1 and the preceding remark, it is

clear that when G is a finite p-group, U is the intersection
of the annihilators of the irreducible kG-modules, and thus is
nothing other than the Jacobson radical of kG . Our next result
is therefore a consequence of Wedderburn's theorem, though it is
easy enough to prove from scratch.

Theorem 2. *When* G *is a finite* p-*group and* k *is the field
of* p *elements, the augmentation ideal* U *of* kG *is the unique
maximal right ideal of* kG *; futhermore,* U *is a nilpotent ideal.*

Proof. A right ideal I of kG is maximal if and only if
kG/I is an irreducible kG-module. Hence, the maximality of U
follows from Remark 4, and if I were another, we would have
kG/I $\cong$ k by Theorem 1. Since I + U = kG , we obtain

$$kG/U \cap I \cong U/U \cap I \oplus I/U \cap I \cong kG/I \oplus kG/U \cong k \oplus k \quad ,$$

so that U = (kG)U $\subseteq$ U $\cap$ I (by Remark 5), and this contradicts
the choice of I $\neq$ U .

To show that U is nilpotent, regard kG = A_o as a right
kG-module, and let A_1 be a maximal submodule (so that A_1 = U
by the above). Let A_2 be a maximal submodule of A_1 and so
on. This process terminates after a finite number of steps
(|kG| is finite), and the result is a chain

$$kG = A_o > A_1 > \ldots > A_m = \{0\} \quad , \tag{2}$$

for some m ϵ N . ((2) is called a *composition series* for A_o ,
and Theorem 1 implies that m = |G| in this case.) Now each
A_i/A_{i+1} , being irreducible is isomorphic to k by Theorem 1
and so, by Remark 5, is annihilated by U . Hence,

$$A_i U \leq A_{i+1} \quad , \quad 0 \leq i \leq m-1 \quad ,$$

and so

180

$$U^i = U^i kG = U^i A_0 \leq A_i \ , \quad 1 \leq i \leq m \ ,$$

and in particular, $U^m = \{0\}$ as required.

Theorem 3 (Burnside Basis Theorem for kG-modules). *Let* G *be a finite* p-*group,* k *the field of* p *elements, and* A $\neq$ 0 *finitely-generated* kG-*module. Then:*

(i) *The set* AU *of all finite sums of elements of the form* au *,* a $\in$ A, u $\in$ U, *is equal to the intersection* $\Phi(A)$ *of all maximal* kG-*submodules of* A *.*

(ii) *A subset* S = $\{a_1, \ldots, a_\ell\}$ *generates* A *over* kG *if and only if the set*

$$S+ = \{a_i + AU \mid 1 \leq i \leq \ell\}$$

spans A/AU *as a* k-*space.*

(iii) *Setting* d(A) = $\dim_k A/AU$ *,* A *has a presentation*

$$\nu \colon F \twoheadrightarrow A \tag{3}$$

with F kG-*free of rank* d(A) *.*

(iv) *In any free presentation* (3) *with* F *of* kG-*rank* d(A) *, we have* Ker $\nu \leq$ FU *.*

Proof.

(i) Let $M \leq A$ be maximal, so that $A/M \cong k$ by Theorem 1, and as above we deduce that $AU \leq M$. Hence $AU \leq \Phi(A)$. For the reverse inequality observe that every element of A/AU is fixed by every element of G , and so A/AU is nothing more than a vector space over k , and every subspace of it is a kG-submodule. Since A is finitely generated over kG , and thus finite dimensional as a k-space, A/AU also has finite k-dimension. Let

$$B = \{b_1, \ldots, b_\ell\}$$

be a basis for A/AU and let A_i/AU be the subspace spanned by

$B \setminus \{b_i\}$. Then each B_i/AU is a maximal subspace (submodule) of A/AU , and

$$\bigcap_{i=1}^{\ell} A_i/AU = \{AU\} \quad .$$

Thus, $AU = \bigcap_{i=1}^{\ell} A_i$, an intersection of maximal submodules of A .

(ii) The necessity of the condition is obvious. For the sufficiency, assume that $S+$ spans A/AU over k and let B be the kG-submodule of A generated by S , so that $B + AU = A$. Now if $B < A$, B lies inside some maximal submodule M of A , and we have

$$A = B + AU \leq M + \Phi(A) = M < A ,$$

by part (i). So $B = A$ as required.

(iii) Letting $\{a_i + AU \mid 1 \leq i \leq d\}$ be a k-basis for A/AU , so that $d = d(A)$, it follows from (ii) that the set $\{a_1, \ldots, a_d\}$ generates A over kG . Thus the mapping

$$\left. \begin{array}{l} F = kG \oplus \ldots \oplus kG \to A \\ \qquad (\gamma_1, \ldots, \gamma_d) \mapsto \sum_{i=1}^{d} a_i \gamma_i \end{array} \right\} ,$$

with F kG-free of rank d , is a kG-epimorphism and will do for ν .

(iv) Take a free presentation of A of the form (3) and consider the composite

$$F \xrightarrow{\nu} A \xrightarrow{\text{nat}} A/AU ,$$

call it θ , say. Now $FU \leq \text{Ker } \theta$, and

$$F/\text{Ker } \theta \cong \text{Im } \theta = A/AU$$

has k-dimension $d = d(A) = \dim_k F/FU$, by hypothesis. Thus
$FU = \text{Ker } \theta$, and it suffices to prove that θ annihilates
$\text{Ker } \nu$, which is already obvious.

Theorem 4. *When* G *is a finite* p-*group and* k *is the field of*
p *elements, there is a* kG-*free resolution*

$$\cdots \quad M_n \xrightarrow{\partial_n} M_{n-1} \to \cdots \to M_1 \xrightarrow{\partial_1} M_o \xrightarrow{\partial_o} k \to 0 \qquad (4)$$

of k *such that* M_n *has* kG-*rank equal to* $\dim_k H^n(G,k)$.

Proof. Before embarking on the proof, note that $H^n(G,k)$ is a
subfactor of $\text{Hom}_{kG}(M_n,k)$ by Remark 3, and the dimension of this
is just the kG-rank of M_n ; thus the resolution (4) is minimal
in the strongest possible sense. We define a resolution inductively as follows. Put $M_o = kG$, $\partial_o = \varepsilon$ in accordance with
Remark 4. Assume inductively that ∂_n is defined for some
$n \geq 0$, and let ∂_{n+1} be the composite

$$M_{n+1} \xrightarrow{\nu_{n+1}} \text{Ker } \partial_n \xrightarrow{\text{inc}} M_n \quad ,$$

where M_{n+1} is kG-free of rank $d(\text{Ker } \partial_n)$ and ν_{n+1} is given
by (3).

In accordance with the remark at the beginning of the proof,
it is sufficient to prove that $\partial_n^* = 0$ for all n , where $*$ is
the operator $\text{Hom}_{kG}(\ ,k)$. To this end, let $\theta \in \text{Hom}_{kG}(M_n,k)$,
$n \geq 1$, so that $\theta\partial_{n+1}^* = \partial_{n+1}\theta$, which is just the composite

$$M_{n+1} \xrightarrow{\nu_{n+1}} \text{Ker } \partial_n \xrightarrow{\text{inc}} M_n \xrightarrow{\theta} k \quad .$$

Since k is G-trivial, $M_n U \leq \text{Ker } \theta$, and

$$\text{Im } \nu_{n+1} = \text{Ker } \partial_n = \text{Ker } \nu_n \leq M_n U \quad ,$$

by applying Theorem 3(iv) to the presentation $\nu_n : M_n \to \text{Ker } \partial_{n-1}$.
Thus, $\theta\partial_{n+1}^* = 0$ and ∂_{n+1}^* is trivial for $n \geq 1$. That $\partial_1^* = 0$

follows in a similar way, since any $\theta \in \text{Hom}_{kG}(M_o, k)$ has $U = \text{Im } \partial_1$ in its kernel. This completes the proof of Theorem 4.

Remark 7.   We know (cf. Exercise 18.5) that each $H^n(G,k)$ is a vector space over k and as such is specified by its dimension. From Example 18.0 (or Theorem 4 above), we see that $\dim_k H^o(G,k) = 1$.   The numbers

$$d(G) = \dim_k H^1(G,k) \quad , \quad r(G) = \dim_k H^2(G,k) \tag{5}$$

will play a vital role in the next section.   Now it follows from Example 18.1, that $H^1(G,k)$ is just the group $\text{Hom}(G,k)$ of homomorphisms from G to (the additive group of) k .

Now as a group, k is cyclic and has order p , so that for any $\theta \in \text{Hom}(G,k)$, both G' and the subgroup G^p , generated by pth powers of members of G , belong to $\text{Ker } \theta$.   Hence,

$$H^1(G,k) \cong \text{Hom}(G,k) \cong \text{Hom}(G/G'G^p, Z_p) \quad ,$$

and since $G/G'G^p$ is elementary abelian (that is, a k-space), we see that $d(G)$ is just $\dim_k G/G'G^p$, the minimal number of generators of G as a group (see Exercise 11).   On the other hand, the construction of the minimal resolution (4) shows that

$$d(G) = d(\text{Ker } \partial_o) = \dim_k U/U^2 \quad . \tag{6}$$

Exercise 12 contains an explicit isomorphism between the groups $G/G'G^p$ and U/U^2 .

EXERCISE 1.   Check the ring axioms for kG .

EXERCISE 2.   If A is a G-module with $pA = 0$, check that extending the G-action by linearity makes A into a kG-module.

EXERCISE 3.   Check that kG and ^pF_o are isomorphic as kG-modules.

184

EXERCISE 4.   Verify that the notions 'G-free mod p' and 'kG-free'
coincide.

EXERCISE 5.   Check that the augmentation mapping $\epsilon: kG \to k$ is
a homomorphism of rings and of kG-modules.

EXERCISE 6.   Verify that $U = \mathrm{Ker}\ \theta$ has the elements
$\{e-x \mid x \in G\backslash\{e\}\}$ as a k-basis.

EXERCISE 7 (Orbit-Stabilizer Theorem).   If a group G acts as a
permutation group (not necessarily faithful) on a set S , we define, for each $s \in S$:

  $L_s = \{sx \mid x \in G\} \subseteq S$,   the *orbit* of s , and

  $G_s = \{x \in G \mid sx = s\} \subseteq G$,   the *stabilizer* of s .

Prove that G_s is a subgroup of G , and that $|G:G_s| = |L_s|$.

EXERCISE 8.   For G a finite p-group, prove that kG has a
unique minimal right ideal and describe it.

EXERCISE 9.   Prove that for any non-trivial finite p-group G ,
the groups $H^n(G,k)$ are all non-trivial.

EXERCISE 10.   Let G be a cyclic group of p-power order. Prove
that $U \le kG$ is a homomorphic image of kG by finding a suitable
kG-generator. Use this fact to write down a minimal kG-free
resolution of k , and compute the groups $H^n(G,k)$.

EXERCISE 11 (Burnside basis theorem for p-groups; cf. Theorem 3
and Remark 7).   Let G be a finite p-group. Using the fact
that $Z(G) \ne E$, prove that every maximal subgroup is normal and
has index p . Defining the *Frattini subgroup* $\Phi(G)$ of G to
be the intersection of the maximal subgroups of G , show that
$\Phi(G) = G'G^p$. Deduce that a subset X generates G if and only
if the cosets $\{\Phi(G)x \mid x \in X\}$ span $G/\Phi(G)$ as a k-space.

EXERCISE 12.   Prove that the mapping

$$\left.\begin{array}{rcl}G/G'G^p & \to & U/U^2 \\ G'G^px & \mapsto & (x-e) + U^2\end{array}\right\}$$

is an isomorphism of groups.

§20. Presentations of group extensions

The process of writing down a presentation for a given extension
in terms of presentations of its two components is embodied in a
fairly elementary piece of folk-lore, whose quantification yields
a twofold harvest in the theory of finite p-groups.

First of all, we forge a vital link between cohomology theory
and the theory of group presentations by deducing that, when G
is a finite p-group and $k = GF(p)$, any presentation of G involves at least $r(G) = \dim_k H^2(G,k)$ relations. As a consequence
of the theorem on minimal resolutions in §19, we then obtain an
innocent-looking exact sequence of kG-modules, which nevertheless
provides the key to the central result (§21) of this chapter.
Secondly, we obtain as a bonus a fairly accurate upper bound for
the number of groups of a given prime-power order.

We begin by crystallizing the folk-lore referred to above.
To this end, let G and A be groups with given presentations

$$G = <X|R> , \quad A = <Y|S> ,$$

and let

$$1 \to A \xrightarrow{\iota} \tilde{G} \xrightarrow{\nu} G \to 1$$

be a fixed extension of G by A . Let

$$\tilde{Y} = \{\tilde{y} = y\iota \mid y \in Y\} ,$$

and let $\tilde{S} = \{\tilde{s} \mid s \in S\}$ be the set of words in $\tilde{Y}$ obtained from

186

S by replacing each y by $\tilde{y}$ wherever it appears. On the other hand, let

$$\tilde{X} = \{\tilde{x} \mid x \in X\}$$

be members of a transversal for Im ι in $\tilde{G}$ such that $\tilde{x}\nu = x$ for all $x \in X$. Furthermore, for each $r \in R$, let $\tilde{r}$ be the word in $\tilde{X}$ obtained from r by replacing each x by $\tilde{x}$. Now ν annihilates each $\tilde{r}$, and so for all $r \in R$,

$$\tilde{r} \in \text{Ker } \nu = \text{Im } \iota \ ,$$

and since Im ι is generated by the set $\tilde{Y}$, each $\tilde{r}$ can be written as a word - say v_r - in the $\tilde{y}$. We put

$$\tilde{R} = \{\tilde{r}v_r^{-1} \mid r \in R\} \quad .$$

Finally, since Im $\iota \lhd G$, each conjugate

$$\tilde{x}^{-1}\tilde{y}\tilde{x}, \ \tilde{x} \in \tilde{X}, \ \tilde{y} \in \tilde{Y},$$

belongs to Im ι , and so is a word - $w_{x,y}$ say - in the $\tilde{y}$. Putting

$$\tilde{T} = \{\tilde{x}^{-1}\tilde{y}\tilde{x}w_{x,y}^{-1} \mid x \in X, y \in Y\} \quad ,$$

we have the following result.

Theorem 1. *The group* $\tilde{G}$ *has a presentation*

$$<\tilde{X}, \ \tilde{Y} \mid \tilde{R}, \ \tilde{S}, \ \tilde{T}> \quad . \tag{1}$$

Proof. Letting D be the group presented by (1), it follows from the fact that all the relations in (1) hold in $\tilde{G}$ that there is a homomorphism

$$\left.\begin{array}{l} \theta: D \to \tilde{G} \\ \quad \tilde{x} \mapsto \tilde{x} \\ \quad \tilde{y} \mapsto \tilde{y} \end{array}\right\} \, ,$$

by the Substitution Test. The restriction of θ to the subgroup $\langle \tilde{Y} \rangle$ of D gives rise to a homomorphism

$$\left.\begin{array}{l} \theta_1: \langle \tilde{Y} \rangle \to \text{Im } \iota \cong A \\ \quad \tilde{y} \longmapsto y \end{array}\right\} \, ,$$

and since the defining relations S of A (with each y replaced by $\tilde{y}$) all hold in $\langle \tilde{Y} \rangle \leq D$, θ_1 must be a bijection. Now the presence of the relations $\tilde{T}$ in (1) means that $\langle \tilde{Y} \rangle$ is a normal subgroup of D, and since $\langle \tilde{Y} \rangle \theta \leq \text{Im } \iota$, θ induces a homomorphism

$$\left.\begin{array}{l} \theta_2: D/\langle \tilde{Y} \rangle \to \tilde{G}/\text{Im } \iota \cong G \\ \quad \langle \tilde{Y} \rangle \tilde{x} \longmapsto x \end{array}\right\} \, .$$

Now the relations R defining G all hold (with x replaced by $\langle \tilde{Y} \rangle \tilde{x}$) in $D/\langle \tilde{Y} \rangle$, so θ_2 must be a bijection. We thus have a commutative diagram

$$
\begin{array}{ccccccccc}
1 & \longrightarrow & A & \overset{\iota}{\longrightarrow} & \tilde{G} & \overset{\nu}{\longrightarrow} & G & \longrightarrow & 1 \\
& & \big\uparrow{\scriptstyle\theta_1} & & \big\uparrow{\scriptstyle\theta} & & \big\uparrow{\scriptstyle\theta_2} & & \\
1 & \longrightarrow & \langle \tilde{Y} \rangle & \overset{\text{inc}}{\longrightarrow} & D & \overset{\text{nat}}{\longrightarrow} & D/\langle \tilde{Y} \rangle & \longrightarrow & 1
\end{array}
$$

with exact rows. Since θ_1 and θ_2 are isomorphisms, it follows as in Remark 17.6 that θ also is an isomorphism. This proves the theorem.

In order to tie this up with the extension theory of §17, we now introduce a few constraints. For the sake of ease of computation, these constraints are somewhat more stringent than is necessary at this stage. We shall assume then that

(i) G is finite,

(ii)   A is finite and cyclic,

(iii) $\tilde{G}$ is a central extension, that is, the G-action on A is trivial.

We shall take fixed finite presentations

$$G = <X|R> \ , \quad A = <c|c^{\ell}> \tag{2}$$

for G and A , where

$$X = \{x_i \mid 1 \le i \le n\} \ , \quad R = \{r_j \mid 1 \le j \le m\} \quad .$$

Let $\tilde{G}$ have the normalized factor set

$$f: G \times G \to A \quad ,$$

so that $\tilde{G}$ is the set of ordered pairs $\{(x,a) \mid x \in G, a \in A\}$, with multiplication

$$(x,a)(y,b) = (xy, ab.(x,y)f)$$

in accordance with Theorem 17.1.

We thus have an extension

$$1 \to A \overset{\iota}{\to} \tilde{G} \overset{\nu}{\to} G \to 1 \ ,$$

with

$$a\iota = (e,a) \ , \quad (x,a)\nu = x$$

for $a \in A$ and $x \in G$. Furthermore,

$$\tilde{c} = (e,c), \quad \tilde{x}_i = (x_i, e)$$

for $1 \le i \le n$. With the above notation, we now have that for all $x \in X$,

$$w_{x,c} = \tilde{c} \quad,$$

since the extension is central. Also, for each r_j in R,

$$v_{r_j} = \tilde{c}^{\ell_j} \quad,$$

where ℓ_j is an integer determined modulo ℓ.

Theorem 1 now asserts that

$$\tilde{G} = \langle \tilde{X}, \tilde{c} \mid \tilde{R}, \tilde{T}, \tilde{c}^{\ell} \rangle \quad, \tag{3}$$

where

$$\tilde{X} = \{\tilde{x}_i \mid 1 \leq i \leq n\} \quad,$$

$$\tilde{R} = \{\tilde{r}_j \tilde{c}^{-\ell_j} \mid 1 \leq j \leq m\} \quad,$$

$$\tilde{T} = \{[\tilde{c}, \tilde{x}_i] \mid 1 \leq i \leq n\} \quad.$$

This all boils down to the fact that the given factor set $f: G \times G \to A$ determines an m-tuple $(\ell_1, \ldots, \ell_m)$ of integers modulo ℓ. Thus, if V_m denotes the additive group of all such m-tuples, we have a function

$$\left. \begin{array}{l} \eta: Z^2(G,A) \to V_m \\[1em] f \mapsto (\ell_1, \ldots, \ell_m) \end{array} \right\} \quad. \tag{4}$$

Theorem 2.

(i) *The function η of (4) is a homomorphism.*

(ii) $\operatorname{Ker} \eta \leq B^2(G,A)$.

(iii) $|H^2(G,A)| \leq \ell^m$.

Proof.

(i) To prove this, we must examine more closely the role played by the factor set f. For a fixed j between 1 and m, let

$$r_j = s_1 \cdots s_t \quad ,$$

be a reduced word in $X \cup X^{-1}$. Writing, for $1 \le i \le n$,

$$(\widetilde{x_i^{-1}}) = \widetilde{x}_i^{-1} = (x_i^{-1}, (x_i, x_i^{-1})f^{-1}) \quad ,$$

we have that

$$
\begin{aligned}
\widetilde{r}_j &= \widetilde{s}_1 \cdots \widetilde{s}_t \\
&= (s_1, e) \cdots (s_t, e)(e, \prod_{i=1}^{t}{}' (s_i^{-1}, s_i)f^{-1}) \\
&= (s_1 \cdots s_t, \prod_{i=1}^{t-1} (s_1 \cdots s_i, s_{i+1})f \cdot \prod_{i=1}^{t}{}' (s_i^{-1}, s_i)f^{-1}) \quad ,
\end{aligned}
$$

using the fact that f is normalized and letting $\prod'$ signify that the product is to be taken only over those s_i which belong to X^{-1} . It follows that ℓ_j is given by

$$c^{\ell_j} = \prod_{i=1}^{t-1} (s_1 \cdots s_i, s_{i+1})f \cdot \prod_{i=1}^{t}{}' (s_i^{-1}, s_i)f^{-1} \quad . \tag{5}$$

The point is that the right-hand side of this equation is linear in f , which proves that η is a homomorphism.

(ii) If all the ℓ_j are 0 , the relators $\widetilde{r}_j$ all hold in $\widetilde{G}$, so that the mapping

$$\left. \begin{aligned} X &\to \widetilde{G} \\ x_i &\mapsto \widetilde{x}_i \end{aligned} \right\}$$

extends to a homomorphism – σ say – from G to $\widetilde{G}$, with the property that $\sigma \nu = 1_G$. That $f \in B(G,A)$ now follows from Remarks 5, 7 and 8 of §17.

(iii) We have:

$$|H^2(G,A)| = \frac{|Z^2(G,A)|}{|B^2(G,A)|} \le \frac{|Z^2(G,A)|}{|\text{Ker } \eta|} = |\text{Im } \eta| \le |V_m| = \ell^m \quad . \tag{6}$$

Remark. It is not too hard to deduce from equation (5) that Ker η = B(G,A) if and only if the exponent-sum of each x_i in each r_j is divisible by ℓ . It follows from the Burnside Basis Theorem (Exercise 19.11) that this condition holds when G is a p-group with X irredundant. Criteria for the second inequality in (6) to be an equality are much harder to find, even for p-groups. Whether every finite p-group has a set of defining relations such that (6) is an equality is an unsolved problem, and forms the subject of §22.

Theorem 3. *Let G be a finite p-group and let*

$$r(G) = \dim_k H^2(G,k) \quad .$$

Then any presentation of G needs at least r(G) relations.

Proof. This is simply a restatement of Theorem 2(iii) with $\ell = p$.

Theorem 4. *For any finite p-group G , there is an exact sequence*

$$A \overset{\alpha}{\twoheadrightarrow} B \overset{\beta}{\twoheadrightarrow} U \to 0$$

of kG-modules, where
(i)   U is the augmentation ideal of kG ,
(ii)   A,B are kG-free, of ranks r(G),d(G) respectively,
(iii)   Im $\alpha \subseteq$ BU .

Proof. Since

$$d(G) = \dim_k H^1(G,k) \quad , \quad r(G) = \dim_k H^2(G,k)$$

192

(see 19.(5)), this sequence is merely a piece of the minimal kG-free resolution of k constructed in Theorem 19.4. That Im $\alpha \subseteq BU$ is a consequence of the construction (part (iv) of Theorem 19.3) and is proved as follows, remember. Since β is onto, so is the composite β' :

$$B \xrightarrow{\beta} U \xrightarrow{\text{nat}} U/U^2 .$$

It is clear that $BU \subseteq \text{Ker } \beta'$, and since

$$\dim_k B/\text{Ker } \beta' = \dim_k U/U^2 = d(G) = \dim_k B/BU ,$$

we must have $BU = \text{Ker } \beta'$. Hence,

$$\text{Im } \alpha = \text{Ker } \beta \subseteq \text{Ker } \beta' = BU ,$$

as required.

Having paved the way for the next section, we conclude this one by finding two bounds of independent interest.

Theorem 5 (J.A. Green). *If G is a p-group of order p^a , then $r(G) \leq a(a+1)/2$.*

Proof. By Theorem 3, it will be sufficient to show that G has a presentation with a generators and $a(a+1)/2$ relations. We prove this by induction on a , the case $a = 1$ being trivial. Assume the result for some $a \geq 1$ and let $\tilde{G}$ be a group of order p^{a+1} . By elementary group theory, $\tilde{G}$ has a central subgroup A of order p , and thus is an extension of $G = \tilde{G}/A$ by A :

$$1 \xrightarrow{} A \xrightarrow{\text{inc}} \tilde{G} \xrightarrow{\text{nat}} G \xrightarrow{} 1 .$$

Taking presentations $<c|c^p>$ and $<x_1,\ldots,x_a \mid r_1,\ldots,r_{a(a+1)/2}>$ for A and G , Theorem 1 yields a presentation for $\tilde{G}$ on $a + 1$ generators and

$$a + 1 + a(a+1)/2 = (a+1)(a+2)/2$$

relations. This proves the claim and hence the theorem.

Theorem 6.   *Up to isomorphism, the number of groups of order* p^a *is at most* $p^{(a^3-a)/6}$.

Proof.   The proof is by induction on a , the result being obvious when $a = 1$. A result from elementary group theory asserts that any finite p-group G has non-trivial centre, whence it follows that G has a central subgroup, A say, of order p . Assuming the result for some $a \geq 1$, let G have order p^{a+1} and consider the extension

$$1 \;\to\; A \;\xrightarrow{\text{inc}}\; G \;\xrightarrow{\text{nat}}\; G/A \;\to\; 1 \;. \tag{7}$$

Now let n_a be the number of groups of order p^a , and recall that equivalent extensions yield isomorphic groups. Since every G of order p^{a+1} is a central extension of the form (7) with $A \cong k$, it follows that

$$n_{a+1} \leq n_a \cdot |H^2(G/A,k)| \quad .$$

Using the inductive hypothesis and the preceding theorem, we have

$$n_{a+1} \leq p^{(a^3-a)/6} \cdot p^{a(a+1)/2}$$

$$= p^{((a+1)^3-(a+1))/6} \;,$$

as required.

EXERCISE 1.   With the notation of the remark following Theorem 2, use formula (5) to show that $\text{Ker } \eta = B^2(G,A)$ if and only if the exponent-sum of each x_i in each r_j is divisible by ℓ .

EXERCISE 2.   Let G be a finite p-group and let

$$1 \rightarrow Z_p \rightarrow \tilde{G} \rightarrow G \rightarrow 1$$

be an arbitrary extension of G by Z_p . Prove that either

$$\tilde{G} \cong G \times Z_p \quad \text{or} \quad d(\tilde{G}) = d(G) \quad .$$

EXERCISE 3.   Let G be a finite p-group and let

$$1 = G_0 < G_1 < \ldots < G_a = G \qquad\qquad (8)$$

be a normal series for G such that each factor G_i/G_{i-1} has order p . The factor G_i/G_{i-1} is said to be complemented if G_i/G_{i-1} has a complement in G/G_{i-1} . Prove that the number of complemented factors in any series of the type (8) is equal to $d(G)$.

EXERCISE 4.   Let $G = \langle X|R \rangle$ be a finite p-group with $|X| = d(G)$. Use Exercise 19.11 to show that the exponent-sum of each generator in each relation is a multiple of p .

EXERCISE 5.   Let $\Phi(G)$ denote the intersection of the maximal subgroups of an arbitrary group G . An element x of G is called a non-generator of G if, whenever $S \subseteq G$ is such that $G = \langle S \cup \{x\} \rangle$, then $G = \langle S \rangle$ already. Prove that the set of non-generators of G is precisely $\Phi(G)$.

EXERCISE 6.   Show that the group

$$\langle x,y,z \mid [x,y] = z^n , [y,z] = x^n , [z,x] = y^n = e \rangle \quad ,$$

where $n \in \mathbb{N}$, is a split extension of Z_n by an abelian group of order $\pm((1+in)^n - 1)((1-in)^n - 1)$, where $i = \sqrt{-1}$.

EXERCISE 7.   Show that the mapping that inverts the generators is an automorphism of the von Dyck group $D(\ell,m,n)$. Prove that the corresponding split extension of Z_2 by $D(\ell,m,n)$ is nothing other than the triangle group $\Delta(\ell,m,n)$.

EXERCISE 8 (see §9). Let θ be the automorphism of the Fibonacci group $F(2,n)$ given by permuting the generators in accordance with the cycle $(1\ 2\ \ldots\ n)$. Use Theorem 1 to find a presentation of the corresponding split extension $E(2,n)$ of Z_n by $F(2,n)$.

EXERCISE 9. Apply Tietze transformations to the result of Exercise 8 to deduce that

$$E(2,n) = <x,y \mid x^2 y = y^2 x,\ y^n = e> \ . \qquad (9)$$

Derive a similar presentation for the corresponding extension $E(r,n)$ of Z_n by $F(r,n)$, $(r \geq 3)$, and describe $E(r,n)^{ab}$

EXERCISE 10. In the group presented in (9), prove that $x^n = e$ when n is even.

EXERCISE 11. If L_n denotes the cyclically-presented group $G_{2n+3}(x_1^{-1}x_2 x_{n+4})$, prove that L_n^{ab} is finite and 3-generated for all $n \geq 0$.

EXERCISE 12. Consider the split extension D_n of Z_{2n+3} by L_n^{ab} induced by the cycle $\theta = (1\ 2\ \ldots\ 2n+3)$. As in Exercises 7,8, find a presentation for D_n , and use Tietze transformations to reduce it to the form

$$<x,y \mid x^n yxy^{n+1},\ y^n xyx^{n+1}> \ .$$

EXERCISE 13. For $r \in Z$, $r \geq 2$, let H_r denote the additive group $\{k/r^\ell \mid k \in Z,\ \ell \in N\}$, so that H_r lies between Z^+ and Q^+ and is not finitely generated. Find a presentation for H_r .

EXERCISE 14.   For $r \in Z$, $r \geq 2$,   consider the group
$G_r = \langle x,y \mid y^{-1}xy = x^r \rangle$ of Exercise 7.2. Prove that G_r' is
isomorphic to the group H_r of the previous exercise.

EXERCISE 15.   Use the exact sequence of Theorem 4 to prove that
$r(G) \geq d(G)$ for any finite p-group G . Use Theorem 3 to
deduce the result of Theorem 6.7 for finite p-groups.

EXERCISE 16.   By adjoining suitable relations, show that the
group

$$G_1 = \langle c_1, c_2, c_3, c_4 \mid c_1^2 = c_2^2 = c_3^2 = c_4^2 = e, c_1 c_2 c_3 c_4 = c_2 c_4 c_1 c_3 = c_4 c_3 c_2 c_1 \rangle$$

has $D_\infty = Z_2 * Z_2$ as a factor group.

EXERCISE 17.   Show that the group G_1 of the previous exercise
has an automorphism of order 5 mapping

$$c_1 \mapsto c_2, \; c_2 \mapsto c_3, \; c_3 \mapsto c_4, \; c_4 \mapsto c_1 c_2 c_3 c_4 \; ,$$

and use Theorem 1 to show that the resulting split extension G_2
of Z_5 by G_1 has a presentation

$$G_2 = \langle c,d \mid c^2 = d^5 = (cd)^5 = (cd^2)^5 = e \rangle \; .$$

EXERCISE 18.   Prove that the group G_2 of the previous exercise
is generated by d and cd^3c .

EXERCISE 19.   Show that there is a homomorphism from the Fibonacci
group $F(2,8)$ onto the group G_2 of the previous exercise, and
deduce that $F(2,8)$ is an infinite group.

§21. The Golod-Šafarevič theorem

Theorem 1 (E.S. Golod and I.R. Šafarevič). *Let* G *be a finite* p-*group with* $d = d(G)$ *and* $r = r(G)$. *Then* $r > d^2/4$.

Proof. The proof is founded on the exactness of the sequence

$$A \overset{\alpha}{\to} B \overset{\beta}{\to} U \to 0 \tag{1}$$

of kG-modules, derived in Theorem 20.4, with A and B kG-free of ranks r and d respectively, and

$$\text{Im } \alpha \leq BU . \tag{2}$$

We proceed in three steps.

(i) For $\ell = 0,1,\ldots,$ we define

$$A_\ell = \{a \in A \mid a\alpha \in BU^\ell\} ,$$

the pre-image of BU^ℓ under α. Since $(A_\ell)\alpha \leq BU^\ell$ and $(BU^\ell)\beta \leq U^{\ell+1}$, we have a sequence

$$0 \to A_\ell/A_{\ell+1} \overset{\alpha_\ell}{\to} BU^\ell/BU^{\ell+1} \overset{\beta_\ell}{\to} U^{\ell+1}/U^{\ell+2} \to 0 , \tag{3}$$

for each $\ell \geq 0$, where α_ℓ, β_ℓ are induced by α, β respectively. It is clear that α_ℓ is one-to-one and that β_ℓ is onto, and we claim that the sequence (3) is in fact exact. To see this, observe that

$$\text{Ker } \beta_\ell = \frac{BU^{\ell+1} + \text{Ker } \beta \cap BU^\ell}{BU^{\ell+1}} ,$$

while

$$\text{Im } \alpha_\ell = \frac{BU^{\ell+1} + \text{Im } \alpha \cap BU^\ell}{BU^{\ell+1}} ,$$

so that the exactness of (1) implies that of (3).

(ii) Letting $U^0 = kG$, define for each $\ell \geq 0$,

198

$$e_\ell = \dim_k A_\ell / A_{\ell+1} \quad , \quad d_\ell = \dim_k U^\ell / U^{\ell+1} \quad ,$$

so that

$$e_0, e_1, \ldots, \quad \text{and} \quad d_0, d_1, \ldots$$

are two sequences of non-negative integers, both eventually zero since B and U are finite-dimensional over k . Note that

$$e_0 = 0 \ , \quad d_0 = 1 \ , \quad d_1 = d \ , \tag{4}$$

using (2), Remark 19.4, 19.(6), respectively. The exactness of (3) implies that for all $\ell \geq 0$,

$$e_\ell + d_{\ell+1} = d d_\ell \quad , \tag{5}$$

since B has kG-rank d . Again using (2),

$$(AU^\ell)\alpha \leq BU^{\ell+1} \quad ,$$

so that

$$AU^\ell \leq A_{\ell+1} \quad ,$$

whence for all $\ell \geq 1$,

$$r(d_0 + \ldots + d_{\ell-1}) \geq e_0 + \ldots + e_\ell \quad . \tag{6}$$

(iii)   Define two polynomials

$$g(t) = \sum_{\ell \geq 0} e_\ell t^\ell \ , \quad f(t) = \sum_{\ell \geq 0} d_\ell t^\ell \quad .$$

Now (5), together with the fact that $d_0 = 1$, implies that for all real t ,

$$tg(t) + (f(t)-1) = dtf(t) \ , \tag{7}$$

while from (6) and (4),

$$\frac{rtf(t)}{1-t} \geq \frac{g(t)}{1-t} \tag{8}$$

for $0 < t < 1$. Eliminating $g(t)$ from (7) and (8), we have

$$(rt^2 - dt + 1)f(t) \geq 1$$

for $0 < t < 1$, and since $f(t)$ is positive in this range,

$$rt^2 - dt + 1 > 0 \tag{9}$$

for $0 < t < 1$. Now the minimum of the quadratic $rt^2 - dt + 1$ occurs at $t = d/2r$, and since G is non-trivial,

$$2r > r \geq d \geq 1 \ ,$$

using the exactness of (1). Thus the minimum value of this quadratic occurs in the interval $(0,1)$, where it takes only positive values by (9). It is thus positive everywhere and so has no real root. Hence the discriminant $d^2 - 4r$ must be negative, which proves the theorem.

Theorem 20.3 now yields the following corollary free of charge.

Theorem 2. *If* G *is a finite p-group with presentation* $<X|R>$ *, then* $|X| \geq d(G)$ *and* $|R| > d(G)^2/4$.

We conclude this section with an example.

Example 1 (A.I. Kostrikin). This example is designed to give some idea of the accuracy of the bound in Theorem 2. We merely state the properties of the groups involved, confining their proofs to the exercises. For $n \in N$, and p a fixed prime, consider the group

$$K_n = \langle X, Y \mid R, S, T \rangle \quad ,$$

where

$$X = \{x_1, \ldots, x_n\} \ , \ Y = \{y_1, \ldots, y_n\} \ ,$$

$$R = \{x_i^p = y_i^p = e \mid 1 \le i \le n\} \ ,$$

$$S = \{[x_i, y_j] = e \mid 1 \le i, j \le n\} \ ,$$

$$T = \{[x_i, x_j] = [y_i, y_j] \mid 1 \le i < j \le n\} \ .$$

It is plain that K_n^{ab} is the direct product of $2n$ copies of Z_p , whence $d = d(K_n) = 2n$. The number of relations involved is given by $r = \frac{3}{2}(n^2 + n)$. Hence,

$$r = \frac{3}{8} d^2 + \frac{3}{4} d \quad ,$$

showing that the bound $r > \frac{1}{4} d^2$ is reasonably accurate, providing we can show that the K_n are all finite p-groups. It follows from the relations S and T that any (left-normed) commutator of weight three in the generators must be the identity. That is, K_n has class at most 2 , or equivalently, $K_n' \le Z(K_n)$. Thus, if a, b are any pair of generators, we know that a commutes with $[a, b] = a^{-1} a^b$, whence a commutes with a^b . Hence,

$$[a, b]^p = (a^{-1} a^b)^p = a^{-p} (a^b)^p = a^{-p} (a^p)^b = e \ ,$$

because of the relations R . It therefore follows from what has been said that K_n' has exponent at most p . Furthermore, since K_n' is generated by the $[a, b]$, $a, b \in X \cup Y$ (since these are all central, no conjugates are involved), it follows that K_n' , and hence K_n , must be finite. A simple computation shows that $|K_n|$ is a divisor of $p^{n(n+3)/2}$, as required.

EXERCISE 1.   We define the lower central series $\{\gamma_n(G) \mid n \in N\}$ of a group G as follows:

$$\gamma_1(G) = G \ , \quad \gamma_{n+1}(G) = \langle [x,y], x \in \gamma_n(G), y \in G \rangle \ , \quad n \geq 1 \ .$$

If there is an $n \in \mathbb{N}$ for which $\gamma_{n+1}(G) = E$, then G is said to be *nilpotent*, and the least such n is called the *class* of G . If G is a group of order p^a (p a prime, $a \in \mathbb{N}$, $a \geq 2$), prove that G is nilpotent of class $\leq a - 1$. For which a is this bound achieved?

EXERCISE 2. If $X, Y \subseteq G$ such that

$$G = \langle X \rangle \ , \quad \gamma_n(G)/\gamma_{n+1}(G) = \langle \gamma_{n+1}(G)y, y \in Y \rangle \ ,$$

prove that $\gamma_{n+1}(G)/\gamma_{n+2}(G)$ is generated by the cosets of $\gamma_{n+2}(G)$ containing the members of the set $\{[y,x] \mid y \in Y, x \in X\}$. Defining the *exponent* of a group to be the l.c.m. of the orders of its elements, prove that (with the usual conventions for ∞) the exponent of each *lower central factor*

$$\gamma_1(G)/\gamma_2(G), \ldots, \gamma_n(G)/\gamma_{n+1}(G), \ldots$$

divides that of its predecessor.

EXERCISE 3 (see Example 1). Prove that $|K_n| = p^{n(n+3)/2}$.

§22. Some minimal presentations

If G is a finite p-group, we have seen that G is minimally generated by

$$d(G) = \dim_k H^1(G,k)$$

elements, where $k = GF(p)$. The minimal number $r'(G)$ of relations needed to define G is achieved in a presentation with $d(G)$ generators (see Exercise 1), and we showed in the previous section that $r'(G)$ is at least as big as

$$r(G) = \dim_k H^2(G,k) \quad .$$

Deciding for a given group G whether or not this bound is achieved is in general very difficult, and the problem of deciding whether the class

$$G_p = \{\text{finite p-groups } G \mid r'(G) = r(G)\}$$

contains all finite p-groups is unsolved.

We shall attempt to give some idea of the extent of the class G_p, and conclude by showing that its subgroup closure contains all finite p-groups when p is odd. Instead of working directly with resolutions, we save space by quoting a couple of results on the multiplicator $M(G)$, whose relevance to the problem is as follows. A straightforward argument using the long exact sequence of cohomology shows that if G is a finite p-group such that $M(G)$ is minimally generated by $m(G)$ elements, then

$$r(G) = d(G) + m(G) \quad . \tag{1}$$

Our first step is to show that G_p is closed under the formation of direct products, from which it will follow that G_p contains all finite abelian p-groups (we already know from Beyl's theorem in §7 that all finite metacyclic p-groups belong to G_p). It follows from a theorem of Schur (see Exercise 3) that

$$m(G \times H) = m(G) + m(H) + d(G)d(H) \quad . \tag{2}$$

Theorem 1. *The class G_p is closed under the formation of direct products.*

Proof. Let $G, H \in G_p$, so that we can write

$$G = \langle X \mid R \rangle \quad , \quad H = \langle Y \mid S \rangle \quad ,$$

with

$$|X| = d(G) , \quad |R| = d(G) + m(G) ,$$

$$|Y| = d(H) , \quad |S| = d(H) + m(H) .$$

Now the presentation of $G \times H$ given in §4 is

$$<X,Y \mid R,S,[X,Y]> ,$$

which has $d(G \times H)$ generators (see Exercise 2), and

$$|R| + |S| + |X|.|Y| = d(G) + m(G) + d(H) + m(H) + d(G)d(H)$$

$$= d(G \times H) + m(G \times H)$$

$$= r(G \times H)$$

relations. Hence, $G \times H \in G_p$ as required.

We now turn to a slightly more complicated construction involving two groups G and H, called their (*standard*) *wreath product*. We assume for convenience that G and H are both finite, and that $|H| = n$, with $H = \{h_1,\ldots,h_n\}$ say. Then in accordance with Cayley's theorem, any element $h \in H$ gives rise to an element $\sigma \in S_n$ given by

$$h_i h = h_{i\sigma} , \quad 1 \le i \le n . \tag{3}$$

We now define the *base group* B to be the direct product of n copies of G, whereupon the wreath product $G \wr H$ is just the split extension of H by B, with H-action given by

$$(g_1,\ldots,g_n)^h = (g_{1\sigma^{-1}},\ldots,g_{n\sigma^{-1}}) ,$$

$h \in H, g_1,\ldots,g_n \in G$. Thus, $G \wr H$ is a group of order $n|G|^n$, and we illustrate its importance by giving some examples (see also Exercises 4,5,6).

Example 1.   If G is abelian, so is the base group B , and the action of H imbues it with the structure of an H—module (§17). If G is actually cyclic of prime order p , the base group is a kG—module, and as such is free of rank one.

Example 2.   When $G = H = Z_2$, it is clear that $G \wr H$ is a non-abelian group of order 8 with at least three involutions $(B = Z_2 \times Z_2)$, whence it must be D_4 . Turning to presentations, we obtain the following more illuminating perspective. Putting

$$G = <a \mid a^2> \quad , \quad H = <x \mid x^2> \quad ,$$

and

$$B = G \times G = <a,b \mid a^2, b^2, (ab)^2> \quad ,$$

the action of G on B is given by

$$a^x = b \, , \quad b^x = a \, ,$$

and Theorem 20.1 yields the presentation

$$G \wr H = <a,b,x \mid a^2 = b^2 = (ab)^2 = e, x^2 = e, a^x = b, b^x = a> \quad .$$

The last relation is superfluous, and the second to last can be used to eliminate b .

Now $x^2 = e$, so b = xax and, on substituting for b , the relation $b^2 = e$ becomes superfluous, while $(ab)^2 = e$ becomes $(ax)^4 = e$. Hence,

$$G \wr H = <a,x \mid a^2 = x^2 = (ax)^4 = e> \quad ,$$

and this is just $D(2,2,4) = D_4$.

Example 3. Consider the set of ordered pairs

$$\Omega_p = \{(i,j) \mid 1 \le i,j \le p\} \ ,$$

where p is a prime, and let

$$\beta_1, \ldots, \beta_p, \theta : \Omega_p \to \Omega_p$$

be the bijections given by:

$$(i,j)\beta_k = \begin{cases} (i,j), j \ne k, \\ (i+1,j), j = k, i < p, \\ (1,j), j = k, \ i = p, \end{cases} \qquad (i,j)\theta = \begin{cases} (i,j+1), j < p, \\ (i,1), j = p. \end{cases}$$

Since the β_k are disjoint p-cycles, they generate in S_{p^2} a subgroup B isomorphic the direct product of p copies of Z_p . This is normalized by the subgroup

$$H = \langle\theta\rangle \cong Z_p \ ,$$

since $\theta^{-1}\beta_k\theta = \beta_{k+1}$ (subscripts mod p). It is not hard to show that $\langle\beta_1, \ldots, \beta_p, \theta\rangle$ is isomorphic to $Z_p \wr Z_p$, and since its order is p^{p+1} , it must be a Sylow p-subgroup of S_{p^2} .

Example 4. Extending the previous example, we define a sequence $\{W_n \mid n \in N\}$ of groups inductively as follows:

$$W_1 = Z_p \ , \quad W_n = W_{n-1} \wr Z_p \ ,$$

where p is a fixed prime. It is the burden of Exercise 7 to show that the W_n are the Sylow p-subgroups of the symmetric groups of p-power degree. By partitioning the set $\{1,2,\ldots,n\}$ in accordance with the p-adic decomposition of n , it can be shown (Exercise 8) that for any $n \in N$, the Sylow p-subgroups of S_n are isomorphic to a direct product of wreath powers (the W_i) of Z_p .

206

Although there is a neat homological argument which yields
both $m(G \wr H)$ and a minimal presentation for $G \wr H$ (in certain
cases) at one blow, we simply quote the value of $m(G \wr H)$ here,
and go on to find the minimal presentation. We impose three
extra conditions on the finite p-groups G and H :

$$G \in G_p \ , \quad d(H) = r'(H) \ , \quad p \text{ is odd.} \tag{4}$$

Note that, since H is finite, we have

$$d(H) = r'(H) \geq r(H) \geq d(H) \ ,$$

so that $H \in G_p$, and also that Z_p satisfies the condition
on H .

Under these conditions, we have that

$$d(G \wr H) = d(G) + d(H) \ ,$$

$$m(G \wr H) = m(G) + \frac{n-1}{2} d(G)^2 \ , \tag{5}$$

where $n = |H|$. The first of these equations comes out in the
wash, while the second is a deeper result. We now construct a
presentation for $G \wr H$ by appealing to Theorem 20.1. Given that

$$G = <X|R> \ , \quad H = <Y|S> \ , \quad |H| = n \ ,$$

we have the following presentation for the base group $B \cong G^{\times n}$
(see §4):

$$B = <X_1,\ldots,X_n \mid R_1,\ldots,R_n,[X_i,X_j]> \ ,$$

where the X_i are disjoint copies of X , R_i is obtained from
R by substituting corresponding elements of X_i , and commutators
$[X_i,X_j]$ are included for all values of i and j such that
$1 \leq i < j \leq n$. Alternatively, X_i may be thought of as the set
of those elements of $G^{\times n}$ having an element of x in the ith

place and all other components equal to e . Now let $Y = \{h_1,\ldots,h_n\}$, and let σ_i be the permutation of $\{1,2,\ldots,n\}$ induced by h_i as in (3), $1 \le i \le n$. Now conjugation by the h_i merely permutes corresponding elements of the X_j among themselves, and we can write

$$X_j^{h_i} = X_{j\sigma_i} \quad , \quad 1 \le j \le n , 1 \le i \le k .$$

These relations specify the action of H on B , and we have the following presentation for $G \wr H$:

$$G \wr H = \langle X_1,\ldots,X_n,Y \mid R_1,\ldots,R_n, \{[X_i,X_j]\}, S, \{X_j^{h_i}X_{j\sigma_i}^{-1}\} \rangle , \qquad (6)$$

the respective ranges of the pairs (i,j) being as indicated above.

We now proceed to prune this presentation using Tietze transformations. First note that $\sigma_1,\ldots,\sigma_k$ generate the regular permutation representation of G , so that the action of G on the sets X_i is certainly transitive. The generators in $X_2,\ldots,X_n$ are thus redundant, and since they are conjugates of corresponding elements of X_1 , the relators $R_2,\ldots,R_n$ can also be omitted. The relators $\{X_j^{h_i} X_{j\sigma_i}^{-1}\}$ merely serve to define the superfluous generators in $X_2,\ldots,X_n$, and so these also disappear. Finally, the commutators $\{[X_i,X_j]\}$ have to be written in the form

$$\{[X_1^{h_i},X_1^{h_j}] \mid 1 \le i < j \le n\} \quad ,$$

reordering and taking inverses if necessary. But these relators are all conjugates of members of the set

$$\{[X_1,X_1^{h_j}] \mid 2 \le j \le n\} \quad , \qquad (7)$$

and so these relators will suffice. Finally, choose a subset $L \subseteq H$ such that $L \,\dot\cup\, L^{-1} = H\backslash\{e\}$, and replace (7) by the set

$$\{[X_1,X_1^\ell] \mid \ell \in L\} \quad ,$$

which is permissible since

$$[X_1, X_1^{\ell}] = E \implies [X_1^{\ell^{-1}}, X_1] = E \implies [X_1, X_1^{\ell^{-1}}] = E \quad .$$

Theorem 2. *If* p *is an odd prime,* $G, H \subset G_p$ *, and* $m(H) = 0$ *, then* $G \wr H \in G_p$ *.*

Proof. Noting that the hypotheses of the theorem are equivalent to the conditions (4), we choose a subset $L \subseteq H$ such that $L \cup L^{-1} = H \backslash \{e\}$ (the disjoint union is possible since p is odd), whence $|L| = \frac{1}{2}(n-1)$. The presentation derived above from (6) has the form

$$G \wr H = \langle X_1, Y \mid R_1, S, \{[X_1, X_1^{\ell}] \mid \ell \in L\} \rangle \quad .$$

By hypothesis, we can write

$$G = \langle X \mid R \rangle , \quad H = \langle Y \mid S \rangle \quad ,$$

where

$$|X| = d(G), \ |R| = d(G) + m(C), \ |Y| = |S| = d(H) \ ,$$

and we have a presentation for $G \wr H$ on $d(G) + d(H)$ generators, and

$$d(G) + m(G) + d(H) + \ell.d(G)^2$$

relations. Since $\ell = \frac{n-1}{2}$, this presentation achieves the values given in (5), proving that $G \wr H \in G_p$, as required.

Theorem 3. *If* p *is an odd prime, then the Sylow* p-*subgroups of all finite symmetric groups belong to* G_p *, and the subgroup closure of* G_p *contains all finite* p-*groups.*

Proof.   Since $Z_p \in G_p$ and has trivial multiplicator, a simple
induction using Theorem 2 shows that all its wreath powers W_n
are also in G_p . That G_p contains the Sylow p-subgroups of
symmetric groups now follows from Theorem 1 and Example 4. The
second assertion now follows using the theorems of Cayley and
Sylow.

EXERCISE 1.   Prove that the minimal number $r'(G)$ of relations
needed to define a finite p-group G is achieved in a presentation on $d(G)$ generators.

EXERCISE 2.   Let G be a finite p-group and H a finite q-group,
where p and q are primes. Prove that $d(G \times H) = d(G) + d(H)$
if and only if p = q .

EXERCISE 3.   Deduce formula (2) from the following theorem of
Schur: for any finite groups G,H ,

$$M(G \times H) \cong M(G) \times M(H) \times (G \otimes H) \quad ,$$

where the tensor product is an abelian group defined as follows.
For cyclic groups we have

$$Z_\ell \otimes Z_m = Z_{(\ell,m)} \quad .$$

For finite abelian groups, we use the Basis Theorem (§6) and the
universal properties

$$G \otimes H = H \otimes G \ , \quad G \otimes (H \times K) = (G \otimes H) \times (G \otimes K) \quad .$$

Finally, all finite groups are covered by defining

$$G \otimes H = G^{ab} \otimes H^{ab} \quad .$$

EXERCISE 4.   Prove that the base group of $Z_p \wr H$ and kH are
isomorphic as kH-modules, where k = GF(p) and p is a prime.

EXERCISE 5. Write down economical presentations for $Z_3 \wr Z_2$ and $Z_2 \wr Z_3$.

EXERCISE 6. Prove that $Z_p \wr Z_p$ is isomorphic to a Sylow p-subgroup of S_{p^2} , and find an economical presentation for it.

EXERCISE 7. Prove that the nth wreath power W_n of Z_p is a Sylow p-subgroup of S_{p^n} , for all $n \in N$.

EXERCISE 8. Prove that any $n \in N$ can be written uniquely in the form

$$n = \sum_{i=k}^{0} a_i p^i , \quad 0 \le a_i \le p-1 ,$$

where i runs from 0 to the integer $k = \max\{j \mid p^j \le n\}$.
By a suitable partitioning of the set $\{1, \ldots, n\}$, deduce that the Sylow p-subgroups of S_n are isomorphic to

$$\mathop{\times}_{i=0}^{k} W_i^{\times a_i} ,$$

a direct product of direct powers of the W_i .

EXERCISE 9. If G and H are finite p-groups, prove that

$$d(G \wr H) = d(G) + d(H) .$$

7 · Small cancellation groups

> The Moving Finger writes; and, having writ,
>
> Moves on: nor all thy Piety nor Wit
>
> Shall lure it back to cancel half a Line,
>
> Nor all thy Tears wash out a Word of it.
>
> (Fitzgerald: The Rubá'iyát of Omar Khayyám)

Given a group $G = \langle X|R \rangle$, suppose that the relators in R are
all cyclically reduced and that R is symmetrized, that is, if r
belongs to R , then so do all cyclic conjugates of r and r^{-1} .
Then G satisfies a small cancellation hypothesis if the amount
of cancelling in forming any product rs $(r,s \in R, r \neq s^{-1}$ in
$F(X))$ is limited in one of various senses to be made precise in
§24. The formulation of these hypotheses is inspired by the
properties of the planar diagrams studied in §23. The latter
boast a degree of intrinsic usefulness, and may be thought of as
portions of the Cayley diagram of G adapted to fit inside R^2 .
The power of the hypotheses derives from Euler's formula for
planar graphs, which explains the innocent but pervasive topological overtones encountered in this branch of combinatorial
group theory.

The conclusions that may be inferred from small cancellation
hypotheses form an interrelated hierarchy of properties such as:

(i)   G is infinite,

(ii)   the torsion elements in G can be classified,

(iii) G is SQ-universal, that is, any countable group can be
       embedded in some factor group of G ,

(iv) G contains a non-cyclic free subgroup, and

(v)   G has soluble word problem, that is, there is an algorithm
       for deciding whether or not any given word in F(X) belongs
       to $\bar{R}$.

We content ourselves with just two applications: a theorem of

D.J. Collins of type (iv) in §25 and a result of type (i) on
Fibonacci groups in §26 where the weakness of the conclusion may
be put down to the fact that the F(r,n) don't quite satisfy the
relevant small cancellation hypotheses.

§23. van Kampen diagrams

We shall be concerned here with finite, connected, planar
graphs, whose edges are oriented and labelled by the generators
of a group $G = <X|R>$. Note that both loops and multiple edges
are allowed, and that faces are simply connected, (by the Jordan
Curve theorem). As in the case of the Cayley diagram of G , we
can assign to any path in such a graph Γ a word in $X^{\pm 1}$, and
hence an element of G , in an obvious way. If the word thus
assigned to the boundary of every face of Γ (for some initial
vertex and some orientation) is a member of R , then Γ is
called a *van Kampen diagram* for $G = <X|R>$. A modicum of in-
tuition suffices to show that the boundary label associated with
any such diagram is a product of conjugates of members of R ,
and is thus equal in G to e . van Kampen diagrams thus provide
a useful means of illustrating the deduction of new relations from
old, as the following examples show.

Example 1.   The diagram of Fig.1 illustrates the fact that the
relation $x^4 = e$ holds in the quaternion group

$<x,y \mid xyx = y, \; yxy = x>$.

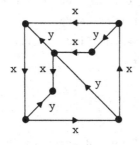

Fig.1

Example 2 (R. Knott). Consider the (by now familiar) group

$$G = \langle x,y \mid x^2yxy^3, y^2xyx^3 \rangle \ .$$

The non-obvious fact that $x^7 = e$ in G is embodied in the following diagram.

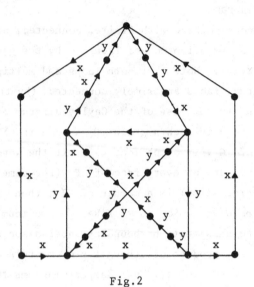

Fig.2

Now let $G = \langle X|R \rangle$ be a presentation in which R is a symmetrized set of cyclically reduced words. This is a standard assumption in small cancellation theory, and any presentation can be augmented to satisfy it in an obvious way without affecting the isomorphism class of the group presented. As pointed out above, the boundary label of any van Kampen diagram for $\langle X|R \rangle$ is a word equal to e in G, and the converse of this assertion forms the starting point of the theory. Thus, if $w \in F(X)$ is a word equal to e in G, we can find $r_1,\ldots,r_n \in R$ and $u_1,\ldots,u_n \in F(X)$ such that

$$w = u_1^{-1}r_1u_1 \cdot u_2^{-1}r_2u_2 \cdot \ \ldots \ \cdot u_n^{-1}r_nu_n \ , \tag{1}$$

and the corresponding van Kampen diagram is just a collection of balloons on strings meeting at a point O, as in the diagram Γ

214

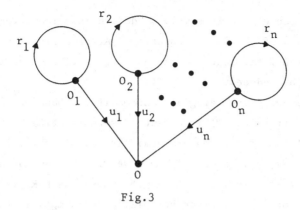

Fig.3

of Fig.3. For each i , $1 \le i \le n$, 0_i is the initial point for the face with boundary label r_i , 0 is the initial point for the boundary of Γ , and the orientation is clockwise in all cases. We thus have a van Kampen diagram for G with boundary label equal to w in $F(X)$.

Now while the boundary label of each face of Γ is reduced, the same is not true for the boundary label of Γ , and we must now modify the diagram to remove this unpleasantness. To this end, assume that the boundary of Γ contains consecutive edges E_1 (from P_1 to P_0) and E_2 (from P_0 to P_2), bearing labels x and x^{-1} , respectively, for some $x \in X^{\pm 1}$. If P_1

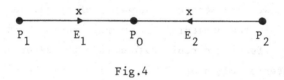

Fig.4

is distinct from both P_0 and P_2 , we can deform that part of Γ lying in a small neighbourhood of E_1 and, keeping P_0 fixed, we can pivot E_1 about P_0 in the exterior of Γ until it comes into coincidence with E_2 . Then P_1 and P_2 will coincide, as will the oriented edges E_1 and E_2 , and we have decreased the length of the boundary of Γ . If P_2 is distinct from both P_0 and P_1 , we proceed in a similar manner, and this leaves us with the case when $P_1 = P_2$. When this occurs, the segment $E_1 E_2$ is a loop attached to the boundary of Γ at the single point P_1 ,

and the procedure here is simply to delete this loop, together
with that part of Γ in its interior. After a finite number of
such operations, the label on the boundary (still equal to w in
F(X)) will be reduced.

We call a van Kampen diagram *reduced* if its boundary label
(for some starting point) is a reduced word in F(X) and, in
addition, it contains no non-trivial circuit whose label is equal
to e in F(X). Thus, for example we wish to exclude such ca-
lamities as that illustrated in Fig.5, where xs ∈ R is in reduced
form. Such contingencies can be obliterated simply by shrinking

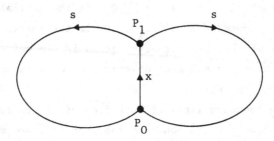

Fig.5

the interior of the offending circuit to a slit, so that the two
faces in Fig.5 are replaced by a simple path from P_1 to P_0
labelled s . We thus perform a sequence of operations of the
type described in the preceding paragraph, operating in the interior of the offending circuit rather than the exterior of Γ .
Thus, only interior edges of Γ are erased, and the boundary of
Γ is unaffected by this procedure. We condense these deliber-
ations into the following fundamental theorem.

Theorem 1.   *If G = <X|R> is a presentation in which R is a
symmetrized set of cyclically reduced words and w is any reduced
word in R̄ , then there is a reduced van Kampen diagram for G
with boundary label w .*

EXERCISE 1.   Convince yourself that the boundary label of any van
Kampen diagram for G = <X|R> is equal in G to the identity.

216

EXERCISE 2.   Let $G = <X|R>$ and suppose that a,b,c,d $\in$ X and [a,c],[a,d],[b,c],[b,d] $\in$ R . Draw the van Kampen diagram for the relator [ab,cd] .

EXERCISE 3.   In the Fibonacci group

$$F(2,4) = <a,b,c,d \mid ab = c , bc = d , cd = a , da = b> ,$$

the relations $a = b^{-3}$, $b^5 = e$ hold; draw the associated van Kampen diagrams.

EXERCISE 4.   Draw a van Kampen diagram to show that the group

$$<a,b \mid abab^2, baba^2>$$

is abelian.

EXERCISE 5.   Given that

$$G = <x,y \mid x^2y = y^2x, x^8 = e> ,$$

show by means of a van Kampen diagram that $y^8 = e$ in G .

EXERCISE 6.   Draw a van Kampen diagram to show that, in the group

$$<x,y \mid x^3 = y^2, x^3 = e, y^{-1}xy = x^2> ,$$

the relation $x^3 = e$ is superfluous.

EXERCISE 7.   Convince yourself that *any* non-trivial circuit whose boundary label is e in F(X) can be eliminated from a van Kampen diagram. (Use induction on the number of faces enclosed by the circuit.)

EXERCISE 8.   If Δ is the Cayley diagram of G = <X|R> and Γ
is a van Kampen diagram for G , prove that there is an incidence
preserving homomorphism of graphs from Γ to Δ .

§24. From Euler's formula to Dehn's algorithm

Having acknowledged the interface between group presentations
and planar graphs, we proceed to establish some fundamental properties of the latter from which we can deduce significant information
about the former. Thus we put group theory aside for the moment
and concentrate on *maps*, that is, finite, connected, planar graphs,
assumed to have more than one vertex to avoid triviality. Though
such a restriction is not necessary at this stage, it is sufficient, from the point of view of applications to van Kampen
diagrams, to consider only maps M such that the boundary D˙
of any face D of M is a simple closed curve.

The *degree* d(D) *of a face* D of M is the number of edges
in its boundary, counted according to multiplicity, and similarly,
the *degree* d(P) *of a vertex* P of M is the number of edges
incident with P . The *boundary* M˙ of M is the topological
boundary of the unbounded component of R^2\M . A *boundary vertex*
or *edge* is one contained in M˙ , and a *boundary face* is one whose
boundary contains a boundary edge; other vertices, edges and faces
are called *interior*.

Our hypotheses on M are roughly to the effect that both the
average degree of a face and the average degree of an interior
vertex are fairly large, that is, at least as big as the corresponding quantities in one of the three regular tesselations of
the plane (cf. §14). Our main conclusions are combinatorial analogues of two elementary metric properties of a sufficiently
regular domain M in R^2 . The first of these asserts that the
integral of curvature along M˙ is a constant (2π) , and the
second bounds the area of M by the square of the length of M˙
times a constant (1/4π) .

In order to formulate the fundamental result, we shall need three functions of a real parameter t. First is the curvature function, defined by

$$K(t) = \sum{}^{\cdot} (t-d(P)) \quad,$$

where the summation ranges over all boundary vertices P. Next, let

$$V(t) = \sum{}^{o} (d(P)-t) \quad,$$

where the summation ranges over all interior vertices, and finally, put

$$R(t) = \sum (d(D)-t) \quad,$$

the sum running over all faces D of M.

Theorem 1. *If M is any map, and p,q are positive real numbers such that $1/p + 1/q = 1/2$, then*

$$K(\tfrac{p}{2}+1) \geq p + V(p) + \tfrac{p}{q} R(q) \quad.$$

Proof. Let v,e,f be the numbers of vertices, edges and faces, respectively, of M, and consider the following three equations:

$$1 = v - e + f \tag{1}$$

$$\sum d(P) = 2e \tag{2}$$

$$\sum d(D) + e_{\infty} = 2e \tag{3}$$

The first of these is Euler's formula, the second is obvious, and in the third, e_∞ is the number of edges on M' . Since the latter are counted according to multiplicity (edges not on the boundary of any face being counted twice), it follows that $e_\infty \geq v_\infty$, the number of boundary vertices.

Letting x be a positive real number, we eliminate e by multiplying equations (1),(2),(3) by $2(x+1),1,x$ respectively, and adding:

$$2(x+1) + \sum d(P) + x\sum d(D) + xe_\infty = 2(x+1)v + 2(x+1)f \quad .$$

Incorporating the terms on the right-hand side into the sums on the left, we obtain

$$2(x+1) + \sum (d(P) - 2(x+1)) + x\sum (d(D) - \frac{2(x+1)}{x}) + xe_\infty = 0 \quad .$$

Now the positive real solutions of $1/p + 1/q = 1/2$ are given parametrically by

$$p = 2(x+1), \quad q = 2(x+1)/x, \quad 0 < x \in R \quad ,$$

and with these substitutions, we have:

$$p + V(p) + \sum^{\cdot}(d(P) - p) + \frac{p}{q} R(q) + \frac{p}{q} e_\infty = 0 \quad ,$$

that is,

$$p + V(p) + \frac{p}{q} R(q) = \sum^{\cdot}(p-d(P)) - \frac{p}{q} e_\infty$$

$$\leq \sum^{\cdot} (p-d(P)) - \frac{p}{q} v_\infty \quad ,$$

since $p/q \geq 0$ and $d_\infty \geq v_\infty$. Now the right-hand side of this inequality is equal to

$$\sum^{\cdot}(p - \frac{p}{q} - d(P)) = K(p - \frac{p}{q}) = K(\frac{p}{2} + 1) \quad ,$$

since $p/2 = 1 + p/q$. This completes the proof.

220

We now turn our hand to the task of deriving some consequences of this result that will later be used to prove group-theoretical theorems. Of the three results stated below, the first follows at once from the fundamental theorem, while the second follows at once from the first, and forms the cornerstone of the next two sections. The third (Theorem 4) also relies on Theorem 2 (this time we supply a proof), and will be used later in this section to solve the word problem for small cancellation groups. At each stage in this progression, something is being thrown away, so that hypotheses can be weakened and conclusions strengthened as indicated, for example, in Exercises 1 and 2.

Theorem 2. *Let* $(p,q) = (6,3)$, $(4,4)$ *or* $(3,6)$ *, and let* M *be a map such that* $d(D) \geq q$ *for each face* D *and* $d(P) \geq p$ *for each interior vertex* P *. Then* $K(p/2 + 1) \geq p$ *, that is*

$$\sum{}^{o}(4-d(P)) \geq 6 \ , \quad \sum{}^{o}(3-d(P)) \geq 4 \ , \quad or \quad \sum{}^{o}(\tfrac{5}{2} - d(P)) \geq 3 \ ,$$

respectively, according to case.

Theorem 3. *If* (p,q) *and* M *satisfy the conditions of the previous theorem, then* M *has at least two boundary vertices of degree* $\leq 3, \leq 2,$ *or* ≤ 2 *, respectively, according to case.*

Theorem 4. *Let* (p,q) *and* M *satisfy the conditions of the previous theorem and define* $a(M)$ *to be the number of vertices of* M *and* $b(M) = K(p)$ *. Then*

$$a(M) \leq \tfrac{1}{J} b(M)^2 \ ,$$

where $J = \min(p^2, 2p(p-2))$ *, that is,* 36,16,6 *respectively, according to case.*

Proof. We begin with a metamathematical remark. The proof will be by induction on $a(M)$, and the inductive step is carried out by passing from M to its double-dual $M_1 = M**$, that is, M_1

is obtained from M by deleting all the vertices on $M^{\cdot}$ together with their incident edges. This lands us squarely in the soup, for the simple reason that M_1 may not be a map; for one thing, it may not be connected, and for another, even its connected components may not be maps. For this reason, we shall prove the theorem for *plans*, that is, finite planar graphs whose connected components are either maps or trivial (that is, have one vertex and no edges). Since every map is a plan, the theorem will follow a fortiori. Needless to say, our plans will satisfy the (p,q)-conditions of Theorem 2, so that M contains no loops or multiple edges, and any vertex of degree two or less must lie on $M^{\cdot}$. The base of our induction thus comes gratis, since the trivial graph has $a = 1$ and $b = p$.

Now let M be a plan with $a(M) \geq 2$, and distinguish two cases. First suppose that M has a boundary vertex of degree less than 2, and let M_0 denote the plan obtained by deleting this vertex along with its incident edge (if it has one). Thus we see that $a_0 = a(M_0) = a-1$, and $b_0 = b(M_0)$ is either $b-p+2$ or $b-p$. Since $b_0^2 \geq J$ and $p \geq 3$, it follows that $b_0 \geq \frac{p+2}{2}$ (see Exercise 3), and, using the inductive hypothesis, we have:

$$b^2 \geq (b_0 + p - 2)^2 = b_0^2 + (p-2)(2b_0 + p - 2)$$

$$\geq Ja_0 + (p-2)(2p) \geq Ja_0 + J = Ja \quad ,$$

as required.

We now turn to the second case, where every boundary vertex of M has degree at least 2 ; it follows that every such vertex is incident with at least two boundary edges. If we let $a^{\cdot}$ denote the number of vertices on $M^{\cdot}$, and write $d^{\cdot} = \sum^{\cdot} d(P)$, we have

$$b = b(M) = K(p) = \sum\nolimits^{\cdot}(p-d(P)) = pa^{\cdot} - d^{\cdot} \quad . \tag{4}$$

Now Theorem 2 asserts that

222

$$p \le K(\frac{p}{2}+1) = \sum\nolimits^{\cdot}(\frac{p}{2}+1-d(P)) = a^{\cdot}(\frac{p}{2}+1) - d^{\cdot} \quad , \tag{5}$$

and, eliminating $d^{\cdot}$ from (4) and (5), we obtain:

$$b \ge \frac{p-2}{2}a^{\cdot} + p \quad . \tag{6}$$

Letting M_1 denote the double dual of M as defined above, we can write

$$d^{\cdot} = f_1 + 2f_2 \quad , \tag{7}$$

where f_i denotes the number of deleted edges with i endpoints on $M^{\cdot}$ $(i = 1,2)$. Furthermore, by the case hypothesis, we also have $d^{\cdot} \ge 2a^{\cdot} + f_1$, whence

$$a^{\cdot} \le f_2 \quad . \tag{8}$$

We proceed to estimate $b_1 = b(M_1) = \sum\nolimits_1^{\cdot}(p-d_1(P))$, where the summation ranges over all vertices P on $M_1^{\cdot}$, and $d_1(P)$ is the degree of P in M_1 . If P is a vertex in M_1 , it cannot be on $M^{\cdot}$, and so $d(P) \ge p$, and if P lies on $M_1^{\cdot}$, we have $d_1(P) = d(P) - f(P)$, where $f(P)$ denotes the number deleted edges incident with P . Since each deleted edge with just one endpoint on $M^{\cdot}$ has its other endpoint on $M_1^{\cdot}$, we see that $\sum\nolimits_1^{\cdot}f(P) = f_1$. Combining these facts, we have:

$$b_1 = \sum\nolimits_1^{\cdot}(p-d_1(P)) \le \sum\nolimits_1^{\cdot}(d(P)-d_1(P)) = \sum\nolimits_1^{\cdot}f(P) = f_1 \quad ,$$

and using (4), we find that

$$b - b_1 \ge pa^{\cdot} - d^{\cdot} - f_1 \quad .$$

Applying (7), (8) and (5) in turn, we have:

$$b - b_1 \ge pa^{\cdot} - 2d^{\cdot} + 2f_2 \ge (p+2)a^{\cdot} - 2d^{\cdot} \ge 2p \quad ,$$

that is,

$$b_1 \leq b - 2p \ . \tag{9}$$

Plugging in the inductive hypothesis that $Ja_1 \leq b_1^2$ (valid even if M_1 is empty), we have

$$Ja = J(a^{\cdot} + a_1) \leq 2p(p-2)a^{\cdot} + b_1^2 \ .$$

Finally, estimating $a^{\cdot}$ and b_1 using (6) and (9) respectively, we have

$$Ja \leq 4p(b-p) + (b-2p)^2 = b^2 \ ,$$

and the proof of Theorem 4 is complete.

Example 1.   We shall prove that the von Dyck group

$$G = D(\ell,m,n) = \langle x,y \mid x^{\ell},y^m,(xy)^n \rangle \ , \quad \ell \geq m \geq n \geq 2 \ ,$$

is infinite whenever $m \geq 4$. Though we have already proved this in Chapter V, the method used here provides a simple and typical application of Theorem 3 above. Note that the definitive result (G infinite if and only if $1/\ell + 1/m + 1/n \leq 1$) is not obtained, and unfortunately this is also typical.

Assuming for a contradiction that G is finite, we know that xy^{-1} has finite order, so there is a $k \in \mathbb{N}$ such that $(xy^{-1})^k$ is the identity in G . It follows from Theorem 23.1 (and Exercise 4) that there is a reduced van Kampen diagram Γ with boundary label $(xy^{-1})^k$. Since $(xy^{-1})^k$ is cyclically reduced, the boundary of Γ contains no vertices of degree 1 . Since each of $x^{\ell},y^m,(xy)^n$ is cyclically reduced, Γ has no interior vertices of degree 1 . Further, one readily observes that, because the three relators are all positive words, Γ has no interior vertices of odd degree (especially 3). Finally, suppose Γ has an interior vertex P of degree 2 . Then P must lie

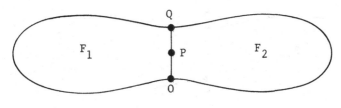

Fig.1

on the boundary of exactly two faces, F_1 and F_2 say, of Γ . If
F_1 and F_2 abut along the segment OPQ , as in Fig.1, the
boundary labels of F_1 and F_2 , reading from $0 \to P \to Q \to 0$ are
freely equal to cyclic conjugates of the defining relators or
their inverses, and have a segment of length two (corresponding to
the two edges incident at P) in common. Inspection reveals that
these two words must in fact be identical, whence the boundary
label of $F_1 \cup F_2$ is freely equal to e , contradicting the fact
that Γ is reduced.

The burden of the previous paragraph is that all the interior
vertices of Γ have degree at least 4 ; that every face has
degree at least 4 is assured by the condition $m \geq 4$. Thus the
hypotheses of Theorem 3 are fulfilled with $(p,q) = (4,4)$, and we
deduce that Γ has a boundary vertex of degree 2 . Letting E_1 and
E_2 be the edges incident at this vertex, the orientations on E_1
and E_2 must be opposite, since the edge–orientations alternate
around the boundary. Further, E_1 and E_2 must lie on the boundary
of a face or else they are traversed twice in opposite directions
in circumnavigating Γ , and $(xy^{-1})^k$ would have to contain x^{-1}
as a letter as well as x . The final contradiction now ensues,
since consecutive edges on any face must have like orientation.

The next job is to translate the threefold hypotheses of
Theorems 2, 3 and 4 into statements about a given presentation
$G = \langle X | R \rangle$. We assume that R is a symmetrized set of cyclically
reduced words (see §23), so that a typical reduced van Kampen
diagram Γ for G can have no interior vertex of degree 1 . We
deal comprehensively with any interior vertex P of degree 2 by
simply deleting it, so that the situation depicted in Fig.1 is

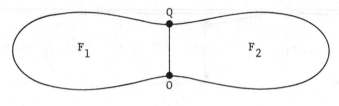

Fig.2

transfigured into that illustrated in Fig.2. Now since Γ is reduced the boundary label of $F_1 \cup F_2$ cannot be freely reducible to e, whence the label on $0Q$ must be an initial segment of two *distinct* members of R.

Definition. Let $G = \langle X|R \rangle$, where R is a symmetrized set of cyclically reduced words. A reduced word w is called a *piece* (with respect to G) if it is an initial segment of two distinct members of R.

It follows that every interior edge of the modified diagram has a piece as its label, and this leads us to formulate the following condition, depending on an integer $k \geq 3$.

$\underline{C(k)}$: $G = \langle X|R \rangle$ satisfies $C(k)$ if no member of R is a product of fewer than k pieces.

It follows that if Γ is a diagram for a $C(k)$-group, then every face of Γ has degree at least k, provided only that every generator is a piece. The last condition is to avoid trouble on the boundary of Γ; the situation where a member of X is not a piece is somewhat pathological, and can usually be handled by other methods, as in §25, for example. The conclusions of Theorems 2, 3 and 4 thus apply to $C(6)$-groups (the case $(p,q) = (3,6)$) whose generators are all pieces.

Passing on to the case $(p,q) = (4,4)$, we wish to preclude interior vertices of degree 3, as illustrated in Fig.3. Letting the boundary labels of F_1, F_2, F_3 (proceeding clockwise from J) be w_1, w_2, w_3, notice that free cancellation occurs in forming each of the products $w_1 w_2$, $w_2 w_3$ and $w_3 w_1$, while no

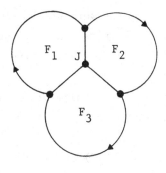

Fig.3

two of w_1, w_2, w_3 can be inverse in $F(X)$. We thus make the
following definition for each integer $k \geq 4$.

$\underline{T(k)}$: $G = \langle X|R \rangle$ satisfies $T(k)$ if for any integer ℓ with
$3 \leq \ell < k$, and any elements $r_1, \ldots, r_\ell \in R$ such that

$$r_1 \neq r_\ell^{-1}, \ r_i \neq r_{i+1}^{-1} , \qquad 1 \leq i \leq \ell-1 ,$$

there is no free cancellation in forming at least one of the
products $r_1 r_2, r_2 r_3, \ldots, r_{\ell-1} r_\ell, r_\ell r_1$.

It is now clear that the conclusions of Theorems 2, 3 and 4
also apply to diagrams of groups satisfying $C(4)$ and $T(4)$ (as
in Example 1), or $C(3)$ and $T(6)$, the condition that each generator is a piece being taken for granted. Of the three small
cancellation conditions

$C(6)$, $C(4)$ and $T(4)$, $C(3)$ and $T(6)$,

we have already had an application of the second, and another is
the topic of the next section. The last is seldom encountered,
as it is often hardest to check, and the conclusion (cf. Theorem 3
above) is often weaker than in the other two cases. So out of
perversity, we give a rare glimpse of this case in §26. We conclude this section by solving the word problem for any finitely-
presented group satisfying any of the three small cancellation
conditions.

Given a group $G = \langle X|R \rangle$, the word problem for G is soluble
if we can decide in a bounded finite number of steps whether any
word $w \in F(X)$ is equal in G to the identity.

Theorem 5. *Let $G = \langle X|R \rangle$ be a finite presentation satisfying
any of the three small cancellation conditions*

$$C(6) \ , \quad C(4) \ and \ T(4) \ , \quad or \quad C(3) \ and \ T(6) \ .$$

Then G has soluble word problem.

Proof. Let $w \in F(X)$ be a reduced word equal to e in G ,
and let Γ be any reduced van Kampen diagram for G with
boundary label w and underlying map M . We claim that the
number $a = a(M)$ of vertices of M is bounded in terms of
$\ell = \ell(w)$. If e_∞, v_∞ denote the number of boundary edges,
boundary vertices of M respectively, we have $v_\infty \leq e_\infty$, as in
the proof of Theorem 1. Furthermore, it is plain that $e_\infty \leq \ell$,
and we see that

$$b(M) = K(p) = \sum{}^{\cdot} (p-d(P))$$

$$\leq \sum{}^{\cdot} p = pv_\infty \leq pe_\infty \leq p\ell \quad .$$

It now follows from Theorem 4 that

$$a \leq \frac{p^2}{J} \, \ell^2 \quad ,$$

where $p^2/J = 1, 1, \frac{3}{2}$ according to case. This establishes the
above claim.

Now let $w \in F(X)$ be any reduced word, with $\ell = \ell(w)$. Now
there are only finitely many maps with $\leq \frac{p^2}{J} \, \ell^2$ vertices, and
each of these has only finitely many edges to be labelled by
generators or pieces from G . Since X and R are finite,
there are only finitely many possible labellings. Now $w = e$
in G if and only if it is the boundary label of one of the

228

resulting finitely many van Kampen diagrams. We thus have an algorithm for solving the word problem for G .

EXERCISE 1.   Prove that the conclusions of Theorems 2, 3, 4 remain valid under the weaker hypotheses that the faces of M have degree at least q and the interior vertices have degree at least p *on the average*, in each of the three cases.

EXERCISE 2.   Let M be the map underlying a reduced van Kampen diagram whose boundary is a simple closed curve. Prove that the inequality in Theorem 1 becomes an equation. Further deduce that, under the conditions of Theorem 3, the number of boundary vertices of degree ≤ 3, ≤ 2 or ≤ 2 is at least 3, 4 or 6, respectively, according to case.

EXERCISE 3.   Let p and b_0 be natural numbers such that

$$p \geq 3 \text{ , and } b_0^2 \geq \min(p^2, 2p(p-2)) \text{ .}$$

Prove that $b_0 \geq (p+2)/2$.

EXERCISE 4.   Write down the symmetrized set of 8 relators obtained from the presentation of the von Dyck group in Example 1. Check that they are cyclically reduced and that the pieces are precisely $x^{\pm 1}, y^{\pm 1}$.

EXERCISE 5.   Investigate the role of a generator x in $G = \langle X | R \rangle$ that is unlucky enough not to be a piece. Specifically, show that if it occurs in any $r \in R$, then r is cyclically conjugate to $(xw)^{\pm k}$, where k is a non-zero integer and w is a word in $X \backslash \{x\}$.

EXERCISE 6.   Note that the word problem for free groups is trivially soluble.

§25. The existence of non-cyclic free subgroups

The method described in the previous section can be exploited to yield more significant information about a group $G = \langle X|R \rangle$ satisfying small cancellation axioms than that a certain element of G has infinite order. Thus, for example, the torsion elements of $C(8)$-groups can be completely classified. Also, under suitable conditions, we can deduce that G has the property indicated by the title of this section, or that G is SQ-universal. The theorem we shall prove has as a kind of prototype the celebrated Freiheitssatz of W. Magnus: if $G = \langle X|r \rangle$ is a one-relator group, there is an $x \in X$ such that the subgroup of G generated by $X \backslash \{x\}$ is free on this set.

Though this theorem can be proved using the methods of small cancellation theory, we choose instead the following result of D.J. Collins: if $G = \langle X|R \rangle$ is a group satisfying $C(4)$ and $T(4)$ and $|X| \geq 3$, then G has a free subgroup of rank 2. It then follows from the Nielsen-Schreier theorem that G has a free subgroup of any given finite rank. In fact, Collins also covers the 2-generator case, where either the desired conclusion holds, or G is one of the groups in a specified finite list. The main body of the proof deals with the case where each generator is a piece, and we embark on this now, consigning the bulk of the opposite case to the exercises.

We assume from now on that $G = \langle X|R \rangle$ satisfies $C(4)$ and $T(4)$, with $|X| \geq 3$ and R a symmetrized set of cyclically reduced relators, and that every member of X is a piece. The force of the small cancellation condition is embodied in the first of two lemmas.

Lemma 1. *Let* $w \in F(X)$ *be a non-empty reduced word equal to* e *in* G. *Then* w *has a subword of length at least two in common with some member of* R.

Proof. Let Γ be a reduced van Kampen diagram for w with underlying map M. We doctor M by deleting any vertex of degree 2 on M that is not on the boundary of a face, replacing its two incident edges by a single edge. Our hypotheses ensure that the resulting map still satisfies the conditions of Theorem 24.3, and since M' contains at most one vertex of degree one, we deduce the existence of a boundary vertex of degree two in M that is also on the boundary of a face. The labels in Γ on the two edges incident with this vertex combine to give the desired common subword.

For tne second lemma, we need the ad hoc notion of a *cancellation triple*, that is, an ordered triple $(xy^{-1}, yz^{-1}, zx^{-1})$ of reduced two-letter words in $F(X)$, $x,y,z \in X^{\pm 1}$.

Lemma 2. *The components of a cancellation triple cannot all be subwords of members of* R .

Proof. Suppose for a contradiction that $(xy^{-1}, yz^{-1}, zx^{-1})$ is a cancellation triple, and that the words

$$u_1 xy^{-1} v_1 \ , \quad u_2 yz^{-1} v_2 \ , \quad u_3 zx^{-1} v_3$$

are in reduced form and belong to R . Since R is symmetrized, it also contains the words

$$x^{-1} v_3 u_3 z \ , \quad z^{-1} v_2 u_2 y \ , \quad y^{-1} v_1 u_1 x \ ,$$

and these violate $T(4)$.

Turning to the proof of the theorem, the action takes place in the complete graph on 6 vertices, to be though of as the 1-skeleton of an octahedron, together with the three diagonals. Specifically, let A,B,C be the points in R^3 whose Cartesian coordinates are $(1,0,0),(0,1,0),(0,0,1)$ respectively, and let A',B',C' be their respective images under reflection in the origin (see Fig.1). If we fix three distinct generators $a,b,c \in X$ then to each of the 30 oriented edges in this figure, we can attach a reduced two-letter word in the following

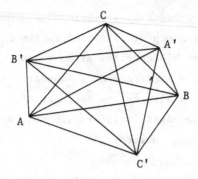

Fig.1

way. The label on an edge leaving (entering) A,B,C,A',B',C'
(ditto) has as its first (second) letter $a^{-1},b^{-1},c^{-1},a,b,c$
$(a,b,c,a^{-1},b^{-1},c^{-1})$ respectively. Thus, if two edges are equal
but for orientation, their labels are inverse to one another. It
is also clear (Exercise 1) that the oriented triangles in this
figure are in one-to-one correspondence with the cancellation
triples involving a, b and c .

We now colour an edge green if its label is a subword of a
member of R , and red otherwise, whereupon Lemma 2 asserts that
there can be no green triangles. The argument now breaks up into
a number of cases, of which the first is typical and given in
full detail now.

1) Let all the edges of the square BCB'C' be red, together
with the diagonal CC' . This means that none of the words

$$b^{-1}c, \ c^{-1}b^{-1}, \ bc^{-1}, \ cb, \ c^2, \ \text{or their inverses,} \tag{1}$$

occur in R (that is, none of them can be a subword of a member
of R). We claim that the subgroup of G generated by
$\{bcb^{-1},c\}$ is free on this set. The reduced form of a supposed
relator is a product of words of the form $bc^m b^{-1},c^n$, with
$mn \neq 0$. The only possible two-letter subwords of this are thus

$$bc, \ c^2, \ cb^{-1}, \ b^{-1}c, \ cb,$$
$$bc^{-1}, \ c^{-2}, \ c^{-1}b^{-1}, \ b^{-1}c^{-1}, \ c^{-1}b,$$

232

and these are precisely the members of the list (1). Thus, no subword of the alleged relator occurs in R , contradicting Lemma 1. It follows that in this case, $\langle bcb^{-1},c\rangle$ is free of rank 2 , as required.

This case forms the first row of the following table, where exactly the same type of argument leads from the red edges in the first column to the free generators in the second (see Exercise 2).

	red edges	free generators
1)	BC,CB',B'C',C'B,CC'	bcb^{-1},c
2)	AB,BC,CA,A'B',B'C',C'A'	$ab^{-1}ca^{-1},bc^{-1}$
3)	A'A,AB,BA',B'C,B'C'	$b^{-1}ab,cb^{-1}abc^{-1}$
4)	AC',C'B,BA,C'B',B'A,A'C	$bacb^{-1},ac$
5)	AB,AB',AC',A'C,CB',B'C',C'B	$c^{-1}a^{-1}bcac,bacb^{-1}$
6)	CC',AB,A'C,CB',B'C',C'B	$bcb^{-1},acbcb^{-1}c^{-1}a^{-1}$
7)	CB',B'A,AC,CA',A'C',C'B	$c^{-1}a^{-1}cbac,bac^{-1}b^{-1}$
8)	CB',B'C',C'B,AB,A'C,C'A,AA',BB'	$a^{-1}bca,acbc^{-1}a^{-1}$
9)	CB',B'C',CC',AC,AC',BA'	$bcb^{-1},a^{-1}ca$

Table 1

From the symmetry of the figure, we can clearly extend our list of arrangements of red edges that yield a pair of free generators. This is done using symmetries that are combinations of the following two basic types.

a) Reflection in a central square: reflection in BCB'C' interchanges A,A' and fixes the other points, whence the triple (a,b,c) is replaced by (a^{-1},b,c) .

b) Rotation through $\pi/2$ about a diagonal: such a rotation about AA' fixes A,A' and maps $B \mapsto C \mapsto B' \mapsto C' \mapsto B$, whereupon (a,b,c) is replaced by (a,c^{-1},b) .

We shall use this symmetry (sometimes implicitly), together with the crucial fact that there are no green triangles, to reduce unknown cases to known cases. At each stage, we preclude the

possibility of cases already dealt with, and so the resulting
stockpile of known cases accumulates as the number of those left
to consider diminishes. When the latter reaches zero, the proof
will be complete. We urge the reader to draw, and suitably colour
in, copies of Fig.1, especially in cases 1)-4) of Table 1. The
proof breaks up into 5 cases, according to the maximum number of
red edges on a central square, and we examine these in turn.

Case IV: there is a central square with four red edges.

By symmetry, there is no loss of generality in assuming that
BCB'C' has four red edges. By 1) and symmetry, both the diagonals
BB' and CC' must be green. Because of 2), at least one of the
eight edges from A or A' to a vertex of BCB'C' must be
green; we can assume that B'A is green. Hence, AB is red and
using 3), either (i) AA' is green, or (ii) BA' is green.

If (i) occurs, B'A' must be red to avoid a green triangle.
By 2) and symmetry, either AC' or A'C is green. In either
case, both AC and A'C' are red, and we have an occurrence
of 2).

Now assume that (ii) occurs and that AA' is red (since (i)
has been dealt with). We again find that B'A' is red and
either AC' or A'C is green. In the former case, AC is red
and by 2), A'C' is green. It follows that A'C is red and we
have an occurrence of 3). When AC' is red and A'C is green,
the argument is virtually identical (interchange A and A',
and C and C').

Case III: there are no central red squares, but there is a
central square with three red edges.

Assume by symmetry that CB',B'C',C'B are red, and that BC
is green. Note that this is the toughest case; we shall invoke
rows 2) - 8) of Table 1. Since there are no green triangles,
either AB or AC is red. Combining two symmetries of types a)
and b) to send (a,b,c) to (a,c,b), we can assume that AB is
red. It follows from 3) that either (i) BA' is green, or (ii)
AA' is green.

If (i) occurs, A'C must be red, and by 6), CC' is green.
Now by 5), either a) AC' is green or b) AB' is green. If a)

234

occurs, AC is red and 3) then makes AA' green. Hence, A'C' is red and 7) forces AB' to be green. It follows that B'A' is red, and we have an occurrence of 4) (triangles B'A'C, B'A'C', and edge AB). Turning to b), we let AB' be green and AC' red. Then B'A' is green by 2), and so both AA' and BB' are red. We are now in case 8) of Table 1.

Passing on to case (ii), we can assume that AA' is green and BA' red. Now 3) forces CC' to be green, and at least one of C'A,C'A is green by 4). Now by a symmetry of type a), we can assume that C'A is green, whence AC and A'C' are red. By 2), AB' is green, so A'B' must be red, and we have an occurrence of 4) (triangles A'C'B',A'C'B , and edge AC).

Case II: There are no central squares with three red edges, but there is one with two red edges.

There are clearly two subcases, according as the red edges on the central square are adjacent or not. We thus have (i) CB' and B'C' red, or (ii) CB' and BC' red.

As usual we take (i) first, and assume that CB' and B'C' are red, while BC and BC' are green. Now CC' must be red, and by 3), either BA or BA' is green. We can assume by symmetry that BA is green, so that AC and AC' are red. Since we are in Case II, A'C and A'C' are both green, whence A'B is red and we are in case 9) of Table 1.

Turning to (ii), we assume that CB' and BC' are red, while BC and B'C' are green. As we are in Case II(ii), not both of AB and AB' are red, and we can assume by symmetry that AB is green. Hence, AC is red and, again since Case II(i) has been dealt with, CA' and AC' are both green. So AB' and BA' are red and since we are in Case II, A'B' is green. It follows that A'C' is red, and we have an occurrence of 2).

Case I: No central square has two red edges, but there is a central square with one red edge.

Suppose that BC is red and that CB', B'C' and C'B are green. Since we are in Case I, either B'A or B'A' is green, and we immediately obtain two red edges in ACA'C' . This contradiction shows that Case I cannot arise.

Case 0: Every central square is green.

This forces all twelve non-diagonal edges in Fig.1 to be green, and so we obtain lots of green triangles, whence this case also cannot arise.

Theorem 1. *Let* $G = \langle X|R \rangle$ *be a presentation with* $|X| \geq 3$ *and* R *a symmetrized set of cyclically reduced words. Suppose that* G *satisfies* C(4) *and* T(4) *and that every generator is a piece. Then* G *contains a non-cyclic free subgroup.*

We close this section with a glance at the situation when not all the generators of $G = \langle X|R \rangle$ are pieces. In view of Exercise 24.5, such generators are of exactly one of the three types covered by the following ad hoc definition.

Definition. A generator is called *singular* if it is not a piece. A singular generator which does not occur in R is called *essential*. A singular generator x that appears only in the cyclic conjugates of

$$(xw)^k , \quad k \in N , \quad w \text{ a word in } X\backslash\{x\} ,$$

is called *removable* if $k = 1$ and a *pole of order* $k-1$ if $k > 1$.

Theorem 2. *Let* $G = \langle X|R \rangle$ *be a finite presentation of a group needing at least three generators. Assume that* R *is a symmetrized set of cyclically reduced words such that* C(4) *and* T(4) *hold. Then* G *contains a non-cyclic free subgroup.*

Proof. Our first step is to use Tietze transformations to remove all the removable generators, noting that this must be done seriatim, since the removal of one generator can affect the status of the others. The result is a presentation such that:

   a)   there are no removable generators (by construction),

   b)   at least three generators are needed (it is still G),

c) $C(4)$ and $T(4)$ still hold (Exercise 3).

If this presentation (still called $<X|R>$ for convenience) has no singular generators, the result follows from Theorem 1. If $<X|R>$ has an essential generator, we have that $G = H * Z$, where H is non-cyclic and Z is infinite cyclic. By Exercise 4, G has a subgroup K of the form $Z * Z$ or $Z_k * Z$ ($k \in N$, $k \geq 2$), and K has a factor group $Z_k * Z_3$, which has a non-cyclic free subgroup by Exercise 12.5. The conclusion for G now follows from Exercise 5. A similar argument for poles shows that G has a subfactor isomorphic to either

$$(Z_2 \times Z_2) * Z_2 \quad \text{or} \quad Z_k * Z_\ell \quad ,$$

where the former is needed to deal with simple poles, and in the latter, k and ℓ are integers ≥ 2 but not both equal to 2 . The result now follows from Exercises 5, 12.5 and 12.6.

EXERCISE 1.   Prove that every reduced two-letter word in the distinct generators a,b,c and their inverses occurs as the label of exactly one edge of Fig.1. By counting cancellation triples and oriented triangles, show that these sets are in one-to-one correspondence.

EXERCISE 2.   Check that in each row of Table 1, the assumption about red edges in the first column yields the free generators in the second column.

EXERCISE 3.   Let $G = <X|R>$ and let $<X'|R'>$ be the result of removing a generator and the corresponding relator (together with its cyclic conjugates and their inverses) by a Tietze transformation. Prove that if $<X|R>$ satisfies $C(k)$ or $T(k)$, then so does $<X'|R'>$.

EXERCISE 4.   Show that every non-cyclic group has a subgroup isomorphic to either Z, Z_k ($k \geq 3$) or $Z_2 \times Z_2$.

EXERCISE 5 (see Exercise 1.5). Let J denote the property of possessing a non-cyclic free subgroup. Prove that if $H \lhd K \leq G$ and K/H has property J , then so does G .

EXERCISE 6. Check that, in the case when G has a pole, the details of the proof of Theorem 2 carry through as stated.

§26. Some infinite Fibonacci groups

We return to the groups $F(r,n) = \langle X|R \rangle$, with

$$X = \{x_1, \ldots, x_n\} \ , \quad R = \{x_{i+1} \cdots x_{i+r} x_{i+r+1}^{-1} \mid 1 \leq i \leq n\} \ ,$$

where subscripts in the relators are reduced modulo n (see §9) . The burning question is whether, for fixed $r \geq 2$ and increasing $n \in N$, the $F(r,n)$ eventually become infinite. We have already shown that though the $F(r,n)^{ab}$ are eventually strictly increasing in size, they are always finite. When r = 2, the only undecided case is when n = 9 , and the bulk of the work for these groups was carried out by R.C. Lyndon, who used small cancellation theoretic methods to prove that $F(2,n)$ is infinite for $n \geq 11$.

We complement his result in this section by showing that $F(r,n)$ is infinite whenever r > 2 and n > 5r . In fact, we obtain the desired result for a slightly larger set of pairs (r,n) than this, and it turns out that the corresponding groups all have soluble word problem. The method provides a rather rare example in the use of conditions C(3) and T(6) , and the overall plan is to discover when $F(r,n)$ satisfies these conditions. The answer turns out to be never (due to the occurrence of the singular vertices described below), and the van Kampen diagrams have to be doctored in order to obtain the desired result. Note that the method breaks down completely when r = 2 , whence our result is disjoint from Lyndon's and is, in fact, an order of magnitude easier to prove.

Our aim is to apply Theorem 24.3 to the map M underlying an

arbitrary reduced van Kampen diagram Γ for $G = F(r,n)$, so we
seek conditions on (r,n) which guarantee that the hypotheses
hold in an amended version of M . As usual, we ignore interior
vertices of degree 2 , and so the investigation begins with the
quest for pieces. Ignoring the trivial case $n = 1$ $(F(r,1) \cong Z_{r-1})$,
we see that every generator is a piece. It is also clear that the

$$x_{i+1} \cdots x_{i+r-1} \; , \quad 1 \leq i \leq n \; , \quad \text{subscripts mod } n \; ,$$

and their subwords are all pieces, and a little thought shows
that these and their inverses are the only ones, except when
$n = 2r$ (a case to be excluded later anyway). Another way of
putting this is to observe that, for $1 \neq n \neq 2r$ and $1 \leq i \leq n$,
any of the three words

$$x_{i+1} \cdots x_{i+r} \; , \quad x_{i+r}^{-1} x_{i+r+1} \; , \quad x_{i+r+1}^{-1} x_{i+1} \tag{1}$$

determines any face of which it forms part of the boundary. Since
Γ is reduced, it follows that no two faces can abut along a path
whose label contains any of these words. We deduce that M
satisfies condition $C(3)$.

Turning to the less straightforward problem of ensuring condition $T(6)$, we attempt to exclude interior vertices of degree
3, 4 or 5 . To this end, we define the *type* of an interior vertex
P to be the sequence of subscripts of the first letters of the
labels on the edges incident at P . We take each in turn, moving
clockwise from a fixed starting point, and indicate that an edge-
label is directed towards P by a vinculum. Now an adjacent pair
of coordinates in such a (cyclically-ordered) $d(P)$-tuple can only
be of one of the following 6 types:

$$(i,i+r) \; , \quad (i,\overline{i-1}) \; , \quad (\overline{i},\overline{i+1}) \; ;$$
$$(i,i-r) \; , \quad (\overline{i},i+1) \; , \quad (\overline{i},\overline{i-1}) \; . \tag{2}$$

Note that by transposing (reflecting about the angle bisector) and then translating, these 6 types correspond in pairs, as indicated by their arrangement in the list (2). An interior vertex P is called *reduced* if, nowhere in its type, does there occur a corresponding pair in adjacent positions. Otherwise, P is called *singular*, so that the type of a singular vertex contains a triple of the form $(1 \leq i \leq n)$

$$(i,i+r,i) \ , \ (i,\overline{i-1},i) \ , \ (\overline{i},\overline{i+1},\overline{i}) \ ,$$

$$(i,i-r,i) \ , \ (\overline{i},i+1,\overline{i}) \ , \ \text{or} \ (\overline{i},\overline{i-1},\overline{i}) \ .$$

We concentrate first on the reduced vertices. By examining the rules (2) for succession of edges, we see (Exercise 2) that for a vertex to have degree 3 , n must be a divisor of one of

$$1, 3, r, 3r \ . \tag{3}$$

Similarly, for degree 4 , n divides one of

$$2, 4, r-1, r+1, 2r, 4r, \tag{4}$$

and for degree 5 , n must divide one of

$$1, 3, 5, r-2, r, r+2, 2r-1, 2r+1, 3r, 5r. \tag{5}$$

Note the presence of the number $r-2$ in (5); it is precisely for this reason that our method fails for the $F(2,n)$.

Passing to singular vertices, note that there can be none of degree 3 , since the boundary labels of faces are reduced. Further, if there were a singular vertex of degree 5 , we could (by identifying the two equal edges and omitting the edge separating them) arrive at the possibility of a vertex of degree 3 , and these have already been dealt with.

There remains the possibility of a singular vertex of degree 4 ; these can actually occur (when $r \geq 3$), but must have the type $(\overline{i},i+1,\overline{i},i+1)$. Our strategy is now to add extra edges

(and possibly interior vertices) to M in such a way that each
such vertex acquires at least 2 new incident edges. Let D be
a face whose boundary contains at least one singular vertex.
Specifically, let there be a singular vertices on the boundary
of D , and put $d(D) = a + b$. Then $a \geq 1$, and the type of a
singular vertex forces $b \geq 2$ (cf. (1)). If $a + b \geq 6$, merely
introduce a new vertex into the interior of D , and join it by
an edge to each vertex on the boundary.

If $a + b < 6$, we have the possibilities

$$(a,b) = (1,2),(2,2),(3,2),(1,3),(1,4),(2,3),$$

corresponding to faces of the types shown in Fig.1. In this
diagram, solid dots denote singular vertices, and there are two
cases when $(a,b) = (2,3)$. The inclusion of the dotted edges
ensures that each singular vertex has its degree increased by

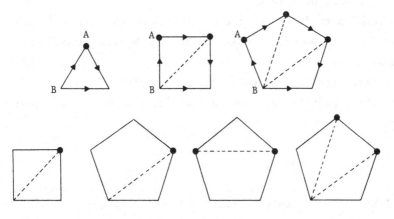

Fig.1

one in each of the last four cases, and similarly for all singular
vertices but A in the first three. The type of a singular vertex ensures that the orientation is as shown, and it remains to
prove that a singular vertex can be of type A on the boundary of
at most two of the four faces that touch it.

If this were not so, such a vertex A would lie on the boundary of two faces, D_1 and D_2 say, abutting along an edge of

type BA . Now the label of this edge is of the form $x_{i+1}\cdots x_{i+\ell}$
(where $1 \le i \le n,\ 1 \le \ell < r$), and so the boundary label of both
D_1 and D_2 (starting at B and moving to A as the next vertex)
is thus $x_{i+1}\cdots x_{i+r}x_{i+r+1}^{-1}$. This shows that the label on the
boundary of $D_1 \cup D_2$ is freely reducible to e , and this contradicts the assumption that Γ is reduced. This proves that the
above construction adds at least two new edges to each singular
vertex, and we conclude that T(6) is fulfilled in the modified
map M' , provided n does not divide any of the numbers in (3),
(4) or (5).

It now follows from Theorem 24.3 that M' has a boundary vertex P of degree at most 3 . Since M' was obtained from M
by *adding* edges (and interior vertices), the same is true for M
and thus for Γ . We shall later choose a positive word for the
boundary label of Γ , and this will force M' to be a simple
closed curve. It follows that each boundary edge of M will be
on the boundary of a face.

It remains to interpret the conclusion of Theorem 24.3, that
is, to list the possible labels on the two boundary edges of Γ
incident at P . Using the rules of succession (2), one readily
checks (Exercise 4) that under the hypothesis of like orientation,
these labels must be x_i followed by x_j , where j is equal to
one of

$$i+1,\ i,\ i+2,\ i+r+1,\ i-r+1, \tag{6}$$

the first being the only possibility when d(P) = 2 .

To complete the proof, assume that G = F(r,n) is finite, so
that for each k between 2 and n , there is a natural number
m such that $(x_1 x_k)^m$ is the identity in G , that is, there is
a reduced van Kampen diagram Γ having this word as its boundary
label. Assuming that n is not a divisor of any of the numbers
in (3), (4) or (5) we have proved that

$$(i,j) = (1,k)\quad \text{or}\quad (k,1) ,$$

where j is one of the numbers listed in (6). This is precluded
by the boundary condition of our theorem, which is now proved in
full.

Theorem 1. *Let* $r \geq 2$ *and* $n \geq 1$ *be integers subject to the
initial condition that* n *is not a divisor of any of*

$$r\pm1, \; r\pm2, \; 2r\pm1, \; 3r, \; 4r, \; 5r,$$

and the boundary condition that the numbers

$$-r, \; -r+2, \; n-1, \; n, \; 1, \; 2, \; 3, \; r, \; r+2$$

do not cover all the residue classes modulo n . *Then* $F(r,n)$ *is
an infinite group.*

EXERCISE 1.   Prove that when $1 \neq n \neq 2r$, the pieces in $F(r,n)$
are just the subwords of $x_{i+1} \cdots x_{i+r-1}$ ($1 \leq i \leq n$) and their
inverses. Find another piece when $n = 2r$.

EXERCISE 2.   Prove that a reduced vertex in a reduced van Kampen
diagram for $F(r,n)$ can have degree 3, 4 or 5 only if n
divides one of the numbers in (3), (4) or (5), respectively.

EXERCISE 3.   Sketch a typical reduced vertex of degree 5 in
the reduced van Kampen diagram for $F(2,n)$.

EXERCISE 4.   Use the rules of succession (2) to show that if
 x_i, x_j are the labels on the boundary edges incident at a vertex
P of degree ≤ 3 , then j is one of the 5 numbers listed in (6).

EXERCISE 5.   Prove that, whatever the value of r , the initial
condition of Theorem 1 forces n to be at least 8 . Show further that when n is 8 or at least 10 , the boundary condition
holds automatically. Deduce that it may be replaced by the simpler
condition that when $n = 9$, $r \not\equiv 4$ or 5 (mod 9) .

EXERCISE 6.   Use Theorem 1 together with the results proved or
quoted in §9 to confirm that, in the problem of deciding when
F(3,n) is infinite, only the cases n = 7, 9 or 15   remain to
be dealt with.

EXERCISE 7.   Prove that when r and n satisfy the conditions
of Theorem 1, F(r,n) has soluble word problem.

EXERCISE 8.   Using the results of Lyndon and Brunner on the
F(2,n) , together with Exercise 9.12, show that when r = kn + 2
(k ∈ N) , then F(r,n) is infinite under either of the conditions:

   n   even and ≥ 6 ,   or k   even and n ≥ 11 .

8 · Groups from topology

Nothing puzzles me more than time and space;
and nothing troubles me less, as I never think
about them. (Lamb: Letter to T. Manning)

As will already be plain to the erudite reader, the connections
between the theory of group presentations and algebraic topology
are both substantial and pervasive. Thus, for example, the triangle groups of Chapter V are essentially geometrical objects,
homological methods play a crucial role in the theory of group
extensions (Chapter VI), and Chapter VII illustrates the depen-
dence of small cancellation methods on properties of planar graphs.
The former subject relies on the latter, both for methods and for
examples and this interrelationship has been increasingly in evi-
dence since the inception of both.

A vital bond, in one direction at least, is forged by the
fundamental group of a space and we begin with a study of this,
carrying with us the idea of a surface for a paradigm. Since the
theory of compact connected n-manifolds is in a sense algebraic-
ally complete when $n = 2$, we go on to study some examples in
the case $n = 3$. We make no apology for our emphasis on algebraic structure and bias towards computational techniques, nor
for the fact that we are merely splashing about on the surface of
what are really very deep waters indeed.

§27. Surfaces

We shall be chiefly concerned with spaces that are locally
Euclidean in the following sense: an n-*manifold* is a Hausdorff
space M in which every point has an open neighbourhood homeomor-
phic to an open subset of $R^n_+ = \{(x_1, \ldots, x_n) \in R^n \mid x_1 \geq 0\}$. The
points of M mapped into $\partial R^n_+ = \{(x_1, \ldots, x_n) \in R^{i1} \mid x_1 = 0\}$ comprise the *boundary* ∂M of M; that this definition is independent

of the choice of open neighbourhood and of homeomorphism is a consequence of Brouwer's theorem on Invariance of Domain.

It is customary to restrict attention to manifolds that are
compact and connected. The first of these strictures is a fairly
specious one, for though an arbitrary manifold can be compactified
by the addition of a single point and the result is a Hausdorff
space (manifolds are obviously locally compact), it may no longer
be locally Euclidean at the extra point. On the other hand, given
that a manifold is compact, the requirement that it be connected
is relatively harmless (see Exercise 2).

We get to grips with the structure of manifolds by means of a
triangulation, and to define this, we need some nomenclature. An
n-*simplex* S is the convex hull of a set V of n+1 affine
independent points in some Euclidean space E (necessarily of
dimension at least n). The convex hull of a subset of V is
called a *face* of S . A finite collection K of simplexes in a
common ambient Euclidean space E is called a *simplicial complex*
if it contains every face of each of its members, and two simplexes
in K intersect in a common face of each. The union of the members of such a K inherits the topology on E and is called the
underlying polyhedron of K , written |K| . Finally a manifold is
triangulable if it is homeomorphic to the underlying polyhedron
of a simplicial complex. It is a profound and fairly recent
result that all manifolds of dimension n are triangulable if
and only if $n \leq 3$.

That 1-manifolds are triangulable is a fairly straightforward
exercise in point-set topology, and given this information, it
is not hard to classify them (see Exercise 3, where the crucial
step is to show that the number of edges meeting at a vertex is
either one or two). Passing on to dimension 2 , we define a
surface to be a compact, connected 2-manifold without boundary.
Our main object in this section is to sketch the classification
of surfaces, and we begin with some familiar examples (see Fig.1).
In each case, the space depicted includes the interior of the
simple closed curve, whose segments are labelled and oriented as
shown. Segments with like labels are identified with one another

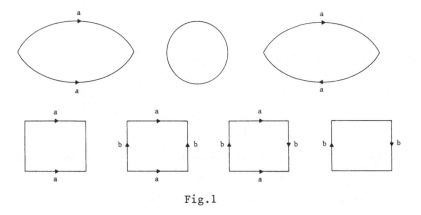

Fig.1

(via the quotient topology), so that the unlabelled segments
comprise the boundary of the resulting surface. With this convention, the diagrams now depict S^2, D, P, C, T, K and M ,
respectively. Only three of these have non-empty boundary,
namely,

$$\partial D = S^1 = \partial M , \quad \partial C = S^1 \mathbin{\dot\cup} S^1 ;$$

we refer to these as *surfaces-with-boundary.*

We skate round the question of orientability by using an
equivalent property: a surface-with-boundary is called *non-
orientable* if and only if it has a subspace homeomorphic to M .
Apart from M itself, exactly two of these 7 spaces are non-
orientable, as the following diagrams show (Fig.2). It is

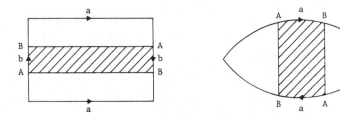

Fig.2

worthwhile to examine what remains in each case when the interior
of the Möbius band is removed: we arrive at the pictures of Fig.3,
where identifications have been made along the dotted lines.

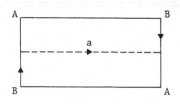 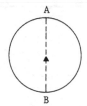

Fig.3

Comparison with Fig.1 shows that this operation turns a Klein bottle (real projective plane) into a Möbius band (closed disc). Turning this process around, we can say that identification of the boundaries (both S^1) of two copies of M yields a Klein bottle, and the same operation performed on a closed disc and a Möbius band yields a projective plane. Alternatively, if we remove the interior of a closed disc from two copies of P and identify the boundaries via a homomorphism, we get a Klein bottle. This suggests the following very useful definition: let M and N be surfaces, remove the interior of a closed disc from each, and identify the boundaries via a homeomorphism. The resulting quotient space is called the *connected sum* of M and N , written M # N .

That this definition is independent of the location of the closed discs is a consequence of the Disc Theorem, and its independence of the choice of identifying homomorphism is shown by similar methods. Assuming this, it is obvious that the operation # is commutative and associative. Furthermore, it is intuitively clear that M # N is orientable if and only if both M and N are.

Given a pair of triangulated surfaces M,N , their connected sum can be formed by removing the interior of a triangle from each and identifying the resulting boundary edges and vertices seriatim in pairs. On the other hand, if M,N are specified as a labelled 2m-,2n-gon (as in Fig.1) respectively, we can proceed as follows. First ensure that the sets $\{x_1,\ldots,x_m\}$ and $\{y_1,\ldots,y_n\}$ of labels are distinct. Then choose an edge of each (labelled x_1,y_1 say) whose end-points are not identified in forming the quotient space; to do this, it may be necessary to break up an

248

edge (and its companion) into two by inserting an extra vertex
at the midpoint and relabelling, as in Fig.4(I) below. Deletion
of the labels x_1 corresponds to removal of an open disc, and
relabelling these edges with y_1 then glueing M and N together along a y_1 yields a picture of M # N as a 2(m+n-1)-gon.
Note that as a special case of this surgery, the reversal of a
single arrow on a labelled 2n-gonal depictation of a surface M
converts it into M # P . This is illustrated by the following
example, which also foreshadows the proof of the classification
theorem for surfaces.

Example 1.   Letting the symbol $\cong$ stand for homeomorphism of
spaces for the nonce, we have seen already that $P \# P \cong K$, and
it is fairly clear that $M \# S^2 \cong M$ for any surface M . It is
much less obvious that $K \# P \cong T \# P$; that this is indeed the
case is illustrated in the following sequence of pictures (Fig.4).

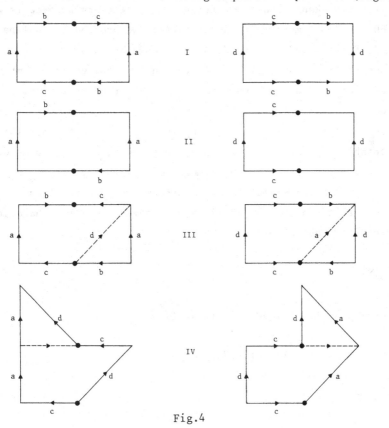

Fig.4

I represents K and T , while in II an open disc has been removed from each. In III, the connected sum is formed with P and a cut made along d,a respectively. IV represents the situation when the identifications corresponding to b are made. The resulting hexagons have pairs of sides identified in accordance with the same code, namely, $a^2 d^{-1} c^{-1} d^{-1} c$, and thus are homeomorphic.

We are now in a position to compile a list of surfaces by applying the operation # to finite collections of copies of the prototypes T and P . Thus, for any integer m ≥ 0 , we define

$$\left. \begin{array}{l} T_m = \text{connected sum of } m \text{ tori } (T_o = S^2) , \\[2mm] P_m = T_m \, \# \, P , \text{ and} \\[2mm] K_m = T_m \, \# \, P \, \# \, P . \end{array} \right\} \tag{*}$$

It is a remarkable and beautiful result that every surface is homeomorphic to onne member of the list (*). We proceed to outline a proof of this, modulo various assumptions that are intuitively reasonable but non-trivial to prove (such as the existence of a triangulation and the invariance of the Euler characteristic - see below).

To see that (*) involves no duplicates, first note that the orientable members of the list are just the T_m . (It turns out that these are precisely the members of (*) that embed in R^3 .) Next, given a triangulation of a surface M that involves f faces, e edges and v vertices, we define the Euler characteristic of M as follows:

$$\chi(M) = v - e + f .$$

Theorem 1. *If M and N are surfaces, then*

$$\chi(M \# N) = \chi(M) + \chi(N) - 2 .$$

Furthermore,

$$\chi(S^2) = 2, \quad \chi(P) = 1, \quad \chi(T) = 0,$$

and

$$\chi(T_m) = 2 - 2m, \quad \chi(P_m) = 1 - 2m, \quad \chi(K_m) = -2m.$$

The surfaces T_m, P_m, K_m *($m \geq 0$)* *are pairwise non-homeomorphic.*

Proof. Assuming M and N to be triangulated, we form their connected sum by removing a face from each and identifying the triangular boundaries as described above. The total number of vertices, edges and triangles thus drops by 3,3,2 respectively. The net diminution of $\chi(M \mathbin{\dot{\cup}} N) = \chi(M) + \chi(N)$ is thus $3 - 3 + 2 = 2$, which proves the first assertion.

Next we compute χ for S^2, P and T, noting that the first of these is just Euler's formula (see formula 24.(1), where we must add 1 to account for the unbounded component of $R^2 \backslash M$). We leave this and P as an exercise, and draw the picture for T (Fig.5). All told, there are 9 vertices (0,A,B,X,Y and the four vertices of the central rectangle), 27 edges (a,b,c,x,y,z and 21 interior ones), and 18 triangles. Thus, $\chi(T) = 9 - 27 + 18 = 0$, as required.

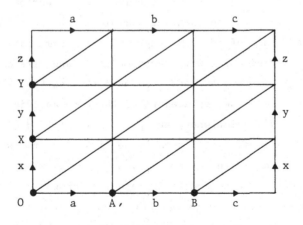

Fig.5

The computation of $\chi(*)$ is a simple induction based on the previous assertion. Finally, given the invariance of χ, the only possible duplication in the list is between T_m and K_{m-1} ($m \in N$), and this is precluded on the grounds of orientability.

Theorem 2. *Every surface is homeomorphic to one of the surfaces* T_m, P_m, K_m ($m \geq 0$) *in the list* (*).

Sketch of proof. Given that an arbitrary surface S is tri-angulable, it follows from properties of simplicial complexes and of the quotient toplogy that S can be represented in the plane by a 2n-gon with labelled oriented edges identified in pairs. Since the order in which the identifications are made is imma-terial, S is determined by the identification code, that is, the word w obtained by juxtaposing the labels on the edges of the 2n-gon. As in §23, this word is determined only up to cyclic conjugacy and inverses.

The identification code for a connected sum of two surfaces is obtained simply by juxtaposing those of the components. Thus, if two adjacent edges of a labelled 2n-gon carry the same label a, the corresponding surface is just the connected sum of a surface with shorter identification code and either S^2 or P, according as the orientations of the edges labelled a is unlike or like. Since (*) contains S^2 and P and is closed under #, we can assume that the 2n-gon representing a minimal counter-example S does not have a pair of adjacent edges with the same label. Now if S has a pair of edges with the same label and like orien-tation, a single surgery (of the type used to pass from III to IV in Fig.4) results in the replacement of these by an adjacent pair. This leaves the situation where any pair of identified edges have opposite orientations. In this case, we first find two such pairs that are separated, as in the code $\ldots x^{-1} \ldots y^{-1} \ldots x \ldots y$, where-upon two surgeries are needed to isolate the commutator $[x,y]$, which then splits off as a torus. The reader is invited to supply the details, or to consult one of the standard texts.

The vital role played in the proof by identification codes

suggests that the following definition might be significant. A group G is called a *surface group* if it has a presentation <X|R> , where R = {r} is a singleton, r is the identification code of a member of (*) , and X consists of the distinct letters involved in r . Specifically, G is a surface group if it has a presentation of the form

$$\langle a_1, b_1, \ldots, a_m, b_m, x_1, \ldots, x_k \mid \prod_{i=1}^{m} [a_i, b_i] \prod_{j=1}^{k} x_j^2 \rangle \quad , \qquad (+)$$

where k = 0 (T_m), 1 (P_m), or 2 (K_m) .

We put this in a broader context by defining a few suitable terms. A continuous mapping f from I = [0,1] into a topological space X is called a *path* from f(0) to f(1) in X . X is said to be *path-connected* if there is a path between any two of its points. The *constant path* at x ∈ X sends everything to x , the *reverse* $\bar{f}$ of the path f sends each t ∈ I to f(1-t) , and the composite of the paths f,g (with f(1) = g(0)) is given by

$$
f.g : I \to X \\
t \mapsto \begin{cases} f(2t) , & t \le \tfrac{1}{2} , \\ g(2t-1) , & t \ge \tfrac{1}{2} . \end{cases}
$$

Furthermore, two paths f,g: I → X are said to be *equivalent* if there is a continuous mapping F: I × I → X such that

$$F(s,0) = f(s) , \quad F(s,1) = g(s) , \quad \text{and}$$

$$F(0,t) = f(0) = g(0) , \quad F(1,t) = f(1) = g(1) .$$

Note that such an f and g must have a common initial point and a common end point, and F is called a *homotopy from* f *to* g *relative to* I' = {0,1} . Finally, a path f: I → X is called a *loop* at x ∈ X if f(0) = f(1) = x .

Bearing these definitions in mind, it is not hard to show that the equivalence classes of loops at x ∈ X form a group under composition of representative paths, and that when X is path-connected, this group is independent of the choice of *base-point* $x \in X$. For a path-connected X, this group is denoted by $\pi_1(X)$, and is called the Poincaré group, first homotopy group, or *fundamental group* of X. The following intuitively reasonable result is stated without proof, and will come as no surprise to the sanguine reader.

Theorem 3.   *The surface group* (+) *is just the fundamental group of the corresponding surface in* (*).

If K is a simplical complex and $|K|$ is connected, we can compute $\pi_1(|K|)$ as the *edge-path group* of K, and there results an alternative method for finding π_1 of a surface. We conclude this section by describing it. Since K is connected, so is its 1-skeleton Γ (set of vertices and edges), and an elementary theorem of graph theory tells us that this contains a maximal tree Δ, that is, Δ is a connected subgraph of Γ having no loops and the same vertex set as Γ. Now orient the edges of Γ and give them distinct labels - call the set of labels X. If R denotes the set of labels on the edges of Δ, and S the set of boundary labels on the triangles of K (so that R consists of one-letter words and S of three-letter words), then the edge-path group of K is given by

$$\pi_1(|K|) = <X|R,S> \quad .$$

Example 2.   We find $\pi_1(T)$ by using the triangulation of Fig.5 and computing its edge-path group. We redraw the picture (Fig.6), indicating the edges of the maximal tree Δ by heavy lines (8 all told, since there are 9 vertices and $\chi(\Delta) = 1$). Note that identified edges receive the same label and that, to save time, we have not labelled or oriented the edges of Δ, since these correspond to the identity element of π_1. As a consequence, the same is true of all edges of all triangles in the shaded region, and hence of the other four unlabelled edges (formerly

254

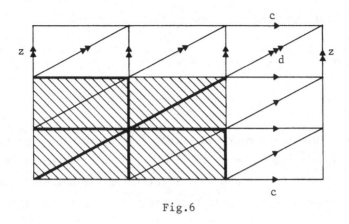

Fig.6

x,y,a,b). Thus the edges $\twoheadrightarrow$ are all z and the edges $\rightarrow$ are all c. Only one extra label (d) is needed, and we obtain the following presentation:

$$\pi_1(T) = <c,d,z \mid zc = d = cz> \quad ,$$

and this is clearly $Z \times Z$, as required.

EXERCISE 1.   Give an example to show that, in the definition of an n-manifold, the requirement that the space be Hausdorff is not superfluous.

EXERCISE 2.   Prove that a compact n-manifold M has only finitely many components, that each of these is a compact n-manifold, and that M is their topological sum.

EXERCISE 3.   Show that a compact connected 1-manifold is homeomorphic either to I or to S^1.

EXERCISE 4.   Use Tietze transformations to prove that

$$<a,b \mid baba^{-1}> \cong <x,y \mid x^2y^2> \quad .$$

[Hint: $P \# P \cong K$.]

EXERCISE 5 (cf. Example 1). Prove that the groups

$$<a,b,c \mid abca^{-1}bc> \ , \ <b,c,d \mid dcbd^{-1}b^{-1}c^{-1}>$$

are isomorphic.

EXERCISE 6. Use the surgery illustrated in Example 1 to prove that the double torus (a pretzel) $T \# T$ can be represented by the decagon of Fig.7. By mentally identifying the pairs of edges v, a, b, c, d respectively, convince yourself that this depictation

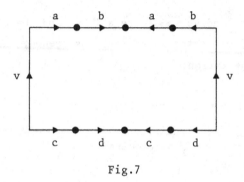

Fig.7

makes sense intuitively. Use Tietze transformations to prove that the groups

$$<a,b,c,d,v \mid vaba^{-1}b^{-1}v^{-1}dcd^{-1}c^{-1}> \ ,$$

$$<x,y,z,t \mid [x,y][z,t]>$$

are isomorphic.

EXERCISE 7. Assuming the classification theorem, convince yourself that $\#$ induces a binary operation on the set of homeomorphism classes of surfaces, and that the resulting monoid has the presentation

$$<t,p \mid tp = pt, \ tp = p^3> \quad .$$

256

EXERCISE 8.   By triangulating S^2 and P , compute their Euler characteristics. (Beware of bogus triangles with only 2 vertices, and of pairs of triangles intersecting in an edge and the other vertex, for these are not allowed.)

EXERCISE 9.   Let M,N be surfaces and remove the interiors of two disjoint closed discs from each, and let M # # N denote the result of identifying their boundaries (both $\cong S^1 \dot{\cup} S^1$). What can you say about this space?

EXERCISE 10.   Given a surface of Euler characteristic χ , triangulated with f triangles, e edges and v vertices, prove that

$$3t = 2e, \quad e = 3(v-\chi), \quad \text{and}$$

$$v \geq \tfrac{1}{2}(7 + \sqrt{(49-24\chi)}) \quad .$$

EXERCISE 11.   Describe surgery that makes two pairs of adjacent like edges adjacent to each other.

EXERCISE 12.   Identify the surface represented by the 2n-gon with boundary label $x_1 x_2 \dots x_n x_1^{-1} \dots x_{n-1}^{-1} x_n$? Do the same for $x_1 x_2 \dots x_n x_1^{-1} x_2^{-1} \dots x_n^{-1}$ (the cases n even or odd are different here).

EXERCISE 13.   Describe a surgical process for reducing a 2n-gonal surface to one in which all vertices become identified in forming the quotient space. [Hint: consider a pair of adjacent vertices in distinct classes A and B , and find a single cut and paste which reduces |B| by 1 , increases |A| by 1 and leaves the other classes fixed.]

EXERCISE 14.   Let G = <X|r> be a one-relator group such that each $x \in X$ appears (as x or x^{-1}) exactly twice in the cyclically reduced word r . By using Tietze transformations corresponding to the surgery employed in Example 1, prove that C is isomorphic to the free product of a free group and a surface group.

EXERCISE 15. Let G be as in the previous exercise. Assuming that the 2n-gon with identification code r has all its vertices identified, prove that G is a surface group (i.e. the free factor is absent). Is the converse true?

EXERCISE 16. Prove that I is connected, and deduce that path-connected spaces are connected.

EXERCISE 17. Use Theorem 3 and the Basis theorem (6.5) to show that a member M of the list (*) is characterised by its *first homology group* $H_1(M) = \pi_1(M)^{ab}$. Observe that in each case, $H_1(M)$ is minimally generated by $2-\chi(M)$ elements.

EXERCISE 18. By triangulating and finding the edge-path group, compute π_1 for S^2, P, K and T_2 .

EXERCISE 19. What can you say about the fundamental group of a surface minus the interiors of one or more closed discs? Experiment with some suitably simple triangulations, modified in the obvious way.

§28. Knots

A *knot* is a continuous injection $\kappa\colon S^1 \to R^3$. Since S^1 is compact and R^3 Hausdorff, such a κ is automatically an embedding. A knot may thus be regarded as an injective loop in R^3 , or as a piece of string in the real world with its ends spliced together. Higher-dimensional knot theory deals with embeddings of S^m in R^n , with $n \geq m+2$ to avoid triviality. We restrict ourselves to the popular case $n = 3 = m+2$, and begin with some examples. The pictures in Fig.1 represent the unknot, the trefoil

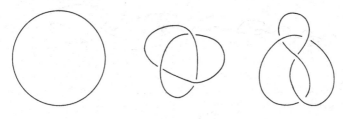

Fig.1

knot, and Listing's knot respectively. We hope these pictures are
self-explanatory, though the idea of planar projection will be
made more precise shortly.

Knot theory is the study of knots up to *equivalence* or *(knot)
type*, where two knots κ and κ' are equivalent if and only if
there is a self-homeomorphism of R^3 sending Im κ to Im κ' .
The *elementary knot deformations* of types 1, 2 and 3 respectively
are illustrated in Fig.2, and it is intuitively clear that local

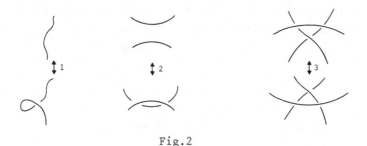

Fig.2

application of each preserves the type of a knot. Though by no
means obvious, the converse is also true: if two knots are equivalent, one can be obtained from the other by a finite sequence of
elementary knot deformations, provided that in each case the
number of crossings (0,3,4 in Fig.1, for example) is finite.

A knot κ is called *polygonal* if Im κ is the union of a
finite set of straight line segments. A knot is called *tame* if
it is equivalent to a polygonal knot and *wild* otherwise. Before
consigning the latter to the shelf of oblivion, we give an
example due to R.H. Fox.

Example 1. In the diagram of Fig.3, the loops decrease in size

Fig.3

and increase in proximity in much the same way as the oscillations
in $f(x) = x \sin 1/x$ as $x \to 0$. The result is a rectifiable
curve and a continuous image of S^1 , just as f becomes continuous at the origin if we define $f(0) = 0$.

The above examples have been portrayed by *planar projection*,
that is, the image of the knot has been projected into a plane in
R^3 by the parallel projection perpendicular to the plane, overcrossings and undercrossings at a double point being indicated in
the obvious way. Such a planar projection is called *regular* if
there result only finitely many multiple points, all these are
double points, and each represents a genuine crossing. Thus we
exclude the pathologies shown in Fig.3 and Fig.4.

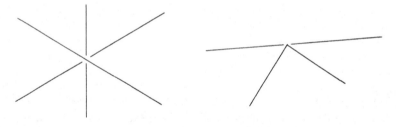

Fig.4

By passing to 3-dimensional projective space, it can be shown
that every polygonal knot has a regular planar projection. Moreover, it turns out that the type of a knot is determined by any
regular planar projection of it. Thus, providing we restrict our
attention to tame knots, a complete theory can (in principle) be
evolved using regular planar projections as models.

A knot κ can be *oriented* by affixing an arrow to Im κ , and
two o-knots are of the same o-*type* if they are equivalent (without

the arrows) via a self-homeomorphism of R^3 that preserves the orientation of their images. Given an o-knot κ , we can form its *reverse* κ_r by reversing the arrow, its *obverse* κ_o by reflecting a regular planar projection in the plane of the paper (overcrossings $\leftrightarrow$ undercrossings), and its *inverse* κ_i by doing both (in either order). A knot κ is *reversible, amphicheiral*, or *invertible* if κ is o-equivalent to κ_r, κ_o or κ_i respectively. For example, Listing's knot is amphicheiral while the trefoil knot is not, and both are reversible. That the resulting tetrachotomy is a genuine one (that is, the four-group acts faithfully on the o-classes) was proved as recently as 1963, when H.F. Trotter exhibited the irreversible knot depicted in Fig.5.

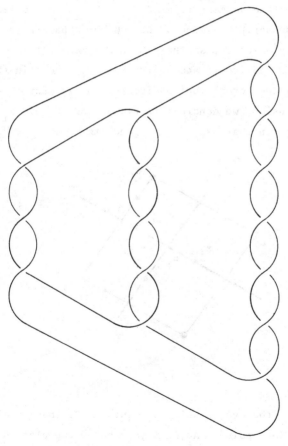

Fig.5

Turning to *invariants* of knots, that is, properties shared by all knots of the same type, we define the *group of a knot* κ to be $\pi_1(R^3 \backslash \text{Im } \kappa)$. We describe how to derive a presentation for this group from a regular planar projection, and also give a method for computing another well-known invariant, the Alexander polynomial. The latter is actually an invariant of the group, and instead of a formal definition, we merely explain a few algorithms for computing it. In fact, the rest of this section is entirely computational.

We begin by deriving the Wirtinger presentation of the group G of a knot κ from a regular planar projection. First orient the knot, and then label the crossings and segments *in order*. Thus, if there are n undercrossings, these are labelled $P_1, \ldots, P_n$ and the segment from P_{i-1} to P_i is labelled x_i ($1 \le i \le n$, subscripts mod n). Let O be a base-point above the plane of projection, so that G has a member that contains a path from O to itself and encircling x_i once (in the sense of a right-handed screw), but encircling no x_j for $j \ne i$. To avoid confusion, we denote this element of G by x_i; it is intuitively plain that $x_1, \ldots, x_n$ generate G.

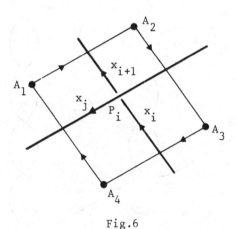

Fig.6

Now suppose the overcrossing segment at P_i has label x_j, and consider four points A_1, A_2, A_3, A_4 below the plane of projection and one in each sector, as indicated in Fig.6. It is clear that the path $0 \to A_1 \to A_2 \to A_3 \to A_4 \to A_1 \to 0$, made up of rectilinear segments, is contractible in the $R^3\backslash \text{Im } \kappa$, and since it is equivalent to

$$0 \to A_1 \to A_2 \to 0 \to A_2 \to A_3 \to 0 \to A_3 \to A_4 \to 0 \to A_4 \to A_1 \to 0 ,$$

the word $x_{i+1}^{-1} x_j x_i x_j^{-1}$ is equal in G to e . If the overhead crossing is from left to right, it is clear that x_j must be re-placed by its inverse. Intuition again asserts (correctly) that these n relators define G . Passing over the unknot, whose group has one generator and no relations, we carry out this com-putation for the trefoil knot.

Example 2. The result of orienting and labelling the knot is depicted in Fig.7, and we easily read off the following

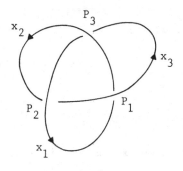

Fig.7

presentation for its group:

$$G = \langle x_1, x_2, x_3 \mid x_2 = x_3^{-1} x_1 x_3, \ x_3 = x_1^{-1} x_2 x_1, \ x_1 = x_2^{-1} x_3 x_2 \rangle .$$

Removing the spare generator x_3 , the two remaining relations coincide, and we have

$$G = \langle x_1, x_2 \mid x_1 x_2 x_1 = x_2 x_1 x_2 \rangle ,$$

or, using further Tietze transformations,

263

$$G = \langle x,y \mid x^2 = y^3 \rangle \quad .$$

In order to derive the Alexander polynomial of a knot from the Wirtinger presentation (or any other) of its group, we need the idea of a Fox derivative. For any word $w \in F = F(X)$, its Fox derivative $\partial w / \partial x$ is defined for each $x \in X$ and belongs to $\mathbb{Z}F$. There is one term of $\partial w / \partial x$ for each occurrence of x in w : if $w = uxv$ (reduced), this term is u , while for $w = ux^{-1}v$, it is $-ux^{-1}$. The sum of these terms is $\partial w / \partial x$. As an example, we express the Fox derivatives of the relators

$$r_1 = x_1 x_3 x_2^{-1} x_3^{-1} \quad ,$$
$$r_2 = x_2 x_1 x_3^{-1} x_1^{-1} \quad ,$$
$$r_3 = x_3 x_2 x_1^{-1} x_2^{-1} \quad ,$$

just obtained for the group of the trefoil knot, as a 3×3 matrix A with (i,j)-entry $\partial r_i / \partial x_j$:

$$A = \begin{pmatrix} 1 & -x_1 x_3 x_2^{-1} & x_1 - x_1 x_3 x_2^{-1} x_3^{-1} \\ x_2 - x_2 x_1 x_3^{-1} x_1^{-1} & 1 & -x_2 x_1 x_3^{-1} \\ -x_3 x_2 x_1^{-1} & x_3 - x_3 x_2 x_1^{-1} x_2^{-1} & 1 \end{pmatrix} \quad .$$

The result of replacing each x_i by t is a matrix over $\mathbb{Z}\langle t \mid \rangle$:

$$\begin{pmatrix} 1 & -t & t-1 \\ t-1 & 1 & -t \\ -t & t-1 & 1 \end{pmatrix} \quad .$$

If we delete the last row and column of this matrix and compute the determinant, the result is the Alexander polynomial of the trefoil knot:

$$A(t) = 1 - t + t^2 \quad .$$

It follows from elementary properties of Fox derivatives that

264

the Alexander polynomial is determined only up to multiplication
by ± an integral power of t . But if we normalize to make it a
polynomial with positive constant term, the result is uniquely
determined by the type of the original knot. There are other
polynomial invariants of knots, but their computation is fairly
complicated, essentially because $Z[t]$ is not a principal ideal
domain.

The Alexander polynomial can also be computed using Seifert's
method, which we now describe. The knot is marked up as above,
and in addition, we attach a sign (±1) to two of the three
segments at each undercrossing as illustrated in Fig.8. When the

Fig.8

overcrossing is from left to right (right to left) the top segment
receives a +1 (-1) , and the segment on its left the other. If
there are n crossings draw up the n×n matrix whose (i,k)-entry
is the label (±1) attached to the segment x_k by the crossing
P_i (0 if it gets no labels, or their sum if it gets more than
one). Delete the last row and column of this matrix, and let S
be the (n-1)×(n-1) matrix whose kth column is the sum of the
first k columns of this. It then turns out that the Alexander
polynomial is just det(S - t(S-I)) , normalized in the usual way.

Example 3.   Marking up Listing's knot according to the above
scheme as in Fig.9, the sequence of three matrices is as follows:

$$
\begin{pmatrix} 1 & 0 & 0 & -1 \\ 1 & 0 & -1 & 0 \\ 0 & -1 & 1 & 0 \\ -1 & 0 & 1 & 0 \end{pmatrix}
\rightarrow
\begin{pmatrix} 1 & 0 & 0 \\ 1 & 0 & -1 \\ 0 & -1 & 1 \end{pmatrix}
\rightarrow
\begin{pmatrix} 1 & 1 & 1 \\ 1 & 1 & 0 \\ 0 & -1 & 0 \end{pmatrix} .
$$

We thus seek the determinant of

$$
\begin{pmatrix} 1 & 1 & 1 \\ 1 & 1 & 0 \\ 0 & -1 & 0 \end{pmatrix} - t \begin{pmatrix} 0 & 1 & 1 \\ 1 & 0 & 0 \\ 0 & -1 & -1 \end{pmatrix} = \begin{pmatrix} 1 & 1-t & 1-t \\ 1-t & 1 & 0 \\ 0 & t-1 & t \end{pmatrix},
$$

and this is easily seen to be $-(t^2 - 3t + 1)$, which may be computed directly, or by first simplifying the matrix using elementary row and column operations. We conclude that $A(t) = 1-3t+t^2$ for Listing's knot.

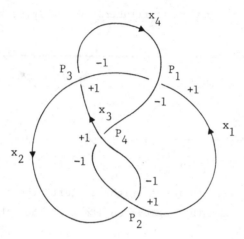

Fig.9

This method can also be used to compute other knot invariants, namely, the invariant factors of the kth homology group of the k-fold cyclic covering manifold associated with the knot, and these are just the invariant factors of the matrix $S^k - (S-I)^k$ for any $k \in N$ (cf. §6).

An alternative method of finding the group of a knot is due to M. Dehn. Orient the knot as above, and label the *faces* (including the outside) by distinct symbols. At each crossing of the type indicated in Fig.10, record the word $ab^{-1}cd^{-1}$. If there are n crossings, the result of putting any face-label equal to the identity will leave n+1 symbols (see Exercise 6) and n words in them; this is a presentation for the group of the knot.

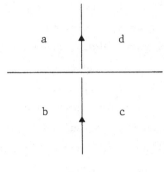

Fig.10

Example 4. The result of orienting Listing's knot and labelling its faces is depicted in Fig.11. Starting at P_1 (see Fig.9), we

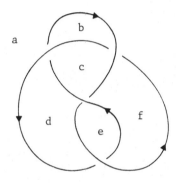

Fig.11

obtain the following relators in turn:

$$cf^{-1}ab^{-1}, \quad ed^{-1}af^{-1}, \quad ad^{-1}cb^{-1}, \quad ef^{-1}cd^{-1} \quad .$$

Letting e fulfil its natural role, the second and fourth of these yield $a = df$ and $c = fd$ respectively. Eliminating these, and also b using the third relator, we are left with

$$\langle d,f \mid fdf^{-1}dfd^{-1}f^{-1}df^{-1}d^{-1}\rangle \quad ,$$

the group of Listing's knot.

By marking up the knot in the same way, we can now describe

Alexander's original method for finding his polynomial. For a
knot with n crossings, draw up the $n \times (n+2)$ matrix whose
columns are headed by the face labels, and whose ith row contains the symbols t, -t, 1, -1 under the labels a, b, c, d
respectively, where the ith crossing is as in Fig.10. Put all
other entries equal to zero, delete any two columns headed by
labels of faces with a common edge, and take the determinant.
This is the Alexander polynomial. Taking Listing's knot as an
example, we obtain the following array:

$$
\begin{pmatrix}
a & b & c & d & e & f \\
1 & -1 & t & 0 & 0 & -t \\
1 & 0 & 0 & -t & t & -1 \\
t & -1 & 1 & -t & 0 & 0 \\
0 & 0 & 1 & -1 & t & -t
\end{pmatrix} .
$$

Deleting the c and d columns, we obtain

$$
\begin{pmatrix}
1 & -1 & 0 & -t \\
1 & 0 & t & -1 \\
t & -1 & 0 & 0 \\
0 & 0 & t & -t
\end{pmatrix} , \rightarrow
\begin{pmatrix}
1 & -1 & 0 & t \\
0 & 1 & t & t-1 \\
0 & t-1 & 0 & t^2 \\
0 & 0 & t & -t
\end{pmatrix} ,
$$

clearing the first column. The determinant is

$$
(t-1)^2 t - (t^3 - t^2(t-1)) = t^3 - 3t^2 + t ,
$$

which normalizes to $1-3t+t^2$, as in Example 3.

We conclude this section with a brief discussion of Conway's
potential function. We need the following definition: a *link* is
an embedding of a finite topological sum of circles in R^3 ,
oriented if its component circles are. Then we can associate
with any oriented knot or link κ a polynomial $\nabla_\kappa(z) \in Z[z]$
in such a way that ∇ of the unknot is the constant polynomial
1 , and the following condition holds. If an overcrossing in κ of

Fig.12

the type shown in the first diagram of Fig.12 is replaced by that
shown in the second, third diagram, we denote the resulting
oriented knot or link by $\bar{\kappa}$, $\overset{o}{\kappa}$, respectively. Then we have

$$\nabla_{\kappa}(z) - \nabla_{\bar{\kappa}}(z) = z\nabla_{\overset{o}{\kappa}}(z) \quad .$$ (*)

It turns out that such a ∇ exists, and is determined uniquely
(not just up to multiplication by $\pm$ a power of z) by the type
of the o-knot or o-link. Furthermore, there is exactly one such
∇ , in that the above three conditions enable us to compute its
value for any o-knot or o-link. Finally, it is related to the
Alexander polynomial A by the following identity:

$$\nabla(x-x^{-1}) = A(x^2) \quad ,$$

suitably normalized.

In order to set up a suitably general example, note that the
diagrams in Fig.12 are the tangles $+1$, -1, 0 of §30. Two tangles
are *added* by splicing together the NE,SE ends of the first to the
NW,SW ends of the second. This operation is plainly associative
and we can add any finite set of tangles together. Given any
tangle, we can splice together its two northern ends, and also its
two southern ends to obtain an o-knot or o-link. For example, if
we add together n copies of the tangle $+1$ ($n \in N$) , we get the

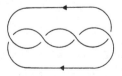
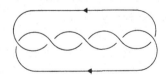

Fig.13

tangle n , whose o-knot or o-link (also called n) is illustrated
in Fig.13 for n = 4,5 . It is clear that 4 is a 2-link and 5 a
5-knot, and that both embed in an (unknotted!) torus in R^3 . We
are now ready to compute the Alexander polynomial of the knot
2n + 1 .

Example 5.   Concentrating on the leftmost crossing (see Fig.13),
it is clear that for n ≥ 2 ,

$$\bar{n} = n - 2 , \quad \overset{o}{n} = n - 1 .$$

It follows from (*) that $\nabla_n(z) = z\nabla_{n-1}(z) + \nabla_{n-2}(z)$ for n ≥ 2 ,
and that $\nabla_o(z) = 0 , \nabla_1(z) = 1$ (see Exercises 9,8). A simple
induction on n now shows that for n ≥ -1 ,

$$\nabla_{n+1}(x-x^{-1}) = \sum_{k=0}^{n} (-1)^k x^{n-2k} ,$$

whence the Alexander polynomial of the torus knot 2m+1 is given
by

$$A(t) = \sum_{i=0}^{2m} (-t)^i .$$

EXERCISE 1.   Use elementary knot deformations to convince yourself that Fig.1 comprises a list of all knots with at most four
crossings. Draw the two five-knots.

EXERCISE 2.   Use the Wirtinger presentation to show that, for
any tame knot κ , $H_1(R^3 \backslash \text{Im } \kappa) \cong Z$.

EXERCISE 3.   Use the previous exercise to show that every knot
group has deficiency 1 .

EXERCISE 4.   Deduce from Exercise 2 that if G is a knot group,
then conjugation within G imbues G'/G'' with the structure of
a ZZ-module.

EXERCISE 5.   Compute the Alexander polynomial of the trefoil knot by Seifert's method, and that of Listing's knot by Fox's method.

EXERCISE 6.   Use Euler's formula to show that the number of faces (including the outside) of a regular planar projection of a knot exceeds the number of its vertices (crossings) by two.

EXERCISE 7.   Compute the group and Alexander polynomial of the trefoil knot using Dehn's method and Alexander's method, respectively.

EXERCISE 8.   Convince yourself that splicing the tangles 1 and 3 yields the unknot and the trefoil knot, respectively.

EXERCISE 9.   A link is called *split* if its component circles can be partitioned into two non-empty sets whose unions are respectively embeddable in disjoint (homeomorphic images of) 3-balls in R^3 . Show that if κ is a split o-link, then $\nabla_\kappa(z) = 0$.

EXERCISE 10.   Use Conway's method to compute the Alexander polynomials of Listing's knot and the *other* 5-knot you drew in Exercise 1 (not the tangle-knot 5).

EXERCISE 11.   Prove that the faces of a knot in regular planar projection can be coloured black and white in such a way that faces with a common edge have opposite colours, as on a chessboard.

EXERCISE 12.   Define the *sum* of two knots κ and κ' to be the result of removing an open line segment ($\cong (0,1)$) from each and identifying the two pairs of boundary points, according to orientation if necessary. Convince yourself that this definition is independent of the location of the deleted segments.

EXERCISE 13. A knot is called *prime* if it admits no decomposition as a sum of non-trivial knots. Show that every tame knot is a sum of prime knots.

EXERCISE 14. Every boy scout knows that there are two inequivalent 6-knots. Draw them, name them, and prove that they have the same group.

EXERCISE 15. Can you distinguish between the true lovers' knot and the false lovers' knot (Fig.14)?

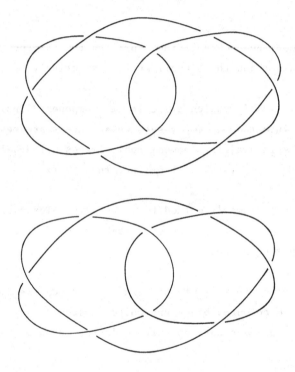

Fig.14

§29. Braids

Consider two parallel planes in R^3 , and name them the *upper* and *lower frame* respectively. For a fixed $n \in N$, choose n distinct points $U_1,\ldots,U_n$ in the upper frame, together with

272

their orthogonal projections $L_1,\ldots,L_n$ in the lower. Now join each U_i to some L_j by a polygonal arc s_i (called the ith *string*) in such a way that:

   (i)   the s_i are pairwise disjoint,

   (ii) any plane between and parallel to the frames meets each s_i exactly once,

   (iii) the correspondence $i \mapsto j$ defines a permutation of $\{1,2,\ldots,n\}$. The resulting configuration is called an n-braid and can be represented in the plane by a suitable parallel projection, in the same fashion as a tame knot. Thus, for example, Fig.1 depicts the 3-braids later to be called x_1 and x_2 .

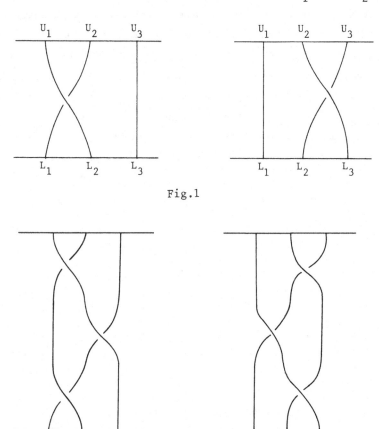

Fig.1

Fig.2

Two n-braids can be composed by hanging the second on to the first, that is, by identifying the lower frame of the latter with

273

the upper frame of the former (in such a way that the chosen
points are identified in the right order), removing this plane,
then compressing the resulting slice of R^3 by an affine transformation to half its thickness. Thus Fig.2 depicts the products
$x_1 x_2 x_1$ and $x_2 x_1 x_2$.

Notice that these two braids are equivalent, in the sense that
an elementary knot deformation of type 3 (see Fig.28.2) transforms
one into the other. Thus two braids are *equivalent* if the planar
projection of one can be deformed into that of the other by a finite sequence of moves of types 2 and 3 (type 1 being precluded
by condition (ii) on the s_i). It is intuitively clear that
composition of classes is independent of choice of representatives
and obeys the associative law. Also, the braid with no crossings
acts as the identity, and the inverse of a braid is obtained by
reflecting it in the lower frame. The resulting group is called
the *braid group on* n *strings* and written B_n .

Now let $\nu: B_n \to S_n$ denote the mapping which assigns to any
braid the permutation $(i \mapsto j)$ of condition (iii) above. Now ν
is an epimorphism and its kernel is called the *unpermuted braid
group*. Further, if S_{n-1} denotes the stabilizer in S_n of 1 ,

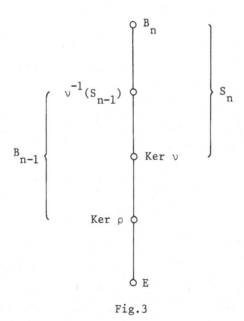

Fig.3

the operation of cutting and removing the first string yields an
epimorphism $\rho: \nu^{-1}(S_{n-1}) \to B_{n-1}$, which preserves the unpermuted
braid group, and whose kernel is called the group of 1-*pure*
braids. We thus arrive at the Hasse diagram of Fig.3. The
obvious inductive value of this chain of subgroups will be
exploited later, at a rigorous level, to elucidate the structure
of the braid groups.

From what has been said, it is clear that B_n is generated by
$\{x_1, \ldots, x_{n-1}\}$, where x_i denotes the braid whose ith string
crosses over its (i+1)st and there are no other crossings. Now
consider the space obtained by removing all n strings of a given
braid from the slice of R^3 between the upper and lower frames.
Its fundamental group is the same as that of $R^2 \backslash \{n$ distinct
points$\}$, namely, the free group F_n of rank n . For free generators, we can take $\{a_1, \ldots, a_n\}$, where a_{k+1} denotes the class
of the loop (based at B to the left of the strings, as in Fig.4)
which passes over the first k+1 strings, under the (k+1)st ,
and then over the first k again back to B . Fig.4 illustrates
the action of x_k on this free generator; 'sliding down the
string' transforms it into $a_{k+1}a_k a_{k+1}^{-1}$. Since the only other
free generator not fixed by x_k is a_k , which passes to a_{k+1} ,
there results an automorphism ξ_k of F_n sending

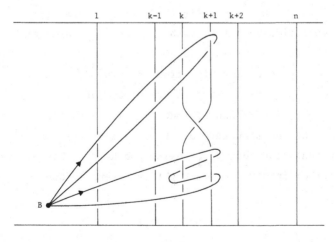

Fig.4

$(a_1, \ldots, a_k, a_{k+1}, \ldots, a_n)$ to the new basis

$$(a_1, \ldots, a_{k+1}, a_{k+1} a_k a_{k+1}^{-1}, \ldots, a_n)$$

(see Exercise 1). With this notation, the *geometrical braid group* on n strings is defined to be the subgroup G_n of Aut F_n generated by $\xi_1, \ldots, \xi_{n-1}$. It will later turn out to be crucial that each element of G_n fixes the product $a_n \ldots a_1$ of free generators.

Now observe that the generators $x_1, \ldots, x_{n-1}$ of the braid group satisfy the relations

$$
\begin{aligned}
S &= \{x_j x_{j+1} x_j = x_{j+1} x_j x_{j+1} \mid 1 \le j \le n-2\} \ , \\
T &= \{x_j x_k = x_k x_j \mid 1 \le j < k-1 < n-1\} \ .
\end{aligned}
\tag{1}
$$

That x_i commutes with x_j when $|i-j| \ge 2$ is clear from the definition, while a typical relation in S is illustrated in Fig.2. We now change the meaning of the symbol B_n , and let it stand for the *algebraical braid group* in n strings, defined as follows:

$$B_n = \langle x_1, \ldots, x_{n-1} \mid S, T \rangle \ .$$

Thus, both G_n and B_n are defined rigorously, and it follows from the Substitution Test that both S_n and G_n are homomorphic images of B_n (see Exercises 2 and 3).

If ν denotes the epimorphism from B_n to S_n fixing the x_i (by san), and S_{n-1} is the stabilizer in S_n of 1 , write $C_{n-1} = \nu^{-1}(S_{n-1})$, a subgroup of index n in B_n . The first step in exploring the structure of B_n , as suggested by Fig.3, is to find a presentation for C_{n-1} using the Reidemeister-Schreier process described in §12. Now a right transversal for S_{n-1} in S_n is given by

$$u_i = x_1 \ldots x_i \ , \quad 0 \le i \le n-1 \ , \tag{2}$$

(cf. the proof of Theorem 5.3), and the corresponding members of B_n thus form a right Schreier transversal for C_{n-1} in B_n . Using the braid relators S and T , together with the fact that $x_i^2 \in \mathrm{Ker}\ \nu \le C_{n-1}$, we see that

$$
\overline{u_i x_j} = \begin{cases}
u_i & , \quad j < i \ , \\
u_{i-1} & , \quad j = i \ , \\
u_{i+1} & , \quad j = i+1 \ , \\
u_i & , \quad j > i+1 \ ,
\end{cases}
\tag{3}
$$

for $0 \le i \le n-1,\ 1 \le j \le n-1$. Thus, C_{n-1} is generated by the $y_{ij} := u_i x_j\ \overline{u_i x_j}^{-1}$, with defining relators

$$
s_{ij} := u_i x_j x_{j+1} x_j^{-1} x_{j+1}^{-1} x_j^{-1} x_{j+1}^{-1} u_i^{-1} \ ,
$$

$$
t_{ijk} := u_i x_j x_k x_j^{-1} x_k^{-1} u_i^{-1} \ ,
$$

with subscript ranges as in (1),(2).

Now the rewriting process is vastly simplified by the identities

$$
y_{ij} y_{i-1j}^{-1} = t_{i-1ij} \ , \qquad 2 \le i+1 < j \le n-1 \ ,
\tag{4}
$$

$$
y_{i-1j} y_{ij}^{-1} = t_{i-1ji} \ , \qquad 2 \le j+1 < i \le n-1 \ ,
\tag{5}
$$

$$
y_{i+1i} y_{i-1i}^{-1} y_{i+1i} = s_{i-1i} \ , \qquad 1 \le i \le n-2 \ ,
\tag{6}
$$

which hold in the free group on $x_1,\ldots,x_{n-1}$. It follows from (3) that, in C_{n-1} ,

$$
y_{ij} = \begin{cases}
x_{j+1} & , \quad j < i \ , \\
a_i & , \quad j = i \ , \\
e & , \quad j = i+1 \ , \\
x_j & , \quad j > i+1 \ ,
\end{cases}
\tag{7}
$$

where $a_i := u_{i-1} x_i^2 u_{i-1}^{-1}$ $(1 \le i \le n-1)$. The result of the re-writing is expressed in Tables 1 and 2 for the s_{ij} and t_{ijk} respectively, where e's appear due to the Tietze transform- ations corresponding to (4),(5),(6).

$j < i-1$	$x_{j+1} x_{j+2} x_{j+1}^{-1} x_{j+2}^{-1} x_{j+1} x_{j+2}^{-1}$	$1 \le j \le n-3$
$j = i-1$	$x_{j+1} a_{j+1} a_j x_{j+1}^{-1} a_j^{-1} a_{j+1}^{-1}$	$1 \le j \le n-2$
$j = i$	$a_j x_{j+1} a_{j+1}^{-1} x_{j+1}^{-1}$	$1 \le j \le n-2$
$j = i+1$	e	(6)
$j > i+1$	$x_j x_{j+1} x_j^{-1} x_{j+1}^{-1} x_j^{-1} x_{j+1}$	$2 \le j \le n-2$

Table 1

$i < j-1$	$x_j x_k x_j^{-1} x_k^{-1}$	$1 < j < k-1 < n-1$
$i = j-1$	e	(4)
$i = j$	$a_j x_k a_j^{-1} x_k^{-1}$	$1 \le j < k-1 < n-1$
$j < i < k-1$	$x_{j+1} x_k x_{j+1}^{-1} x_k^{-1}$	$1 \le j < k-2 < n-2$
$i = k-1$	e	(5)
$i = k$	$x_{j+1} a_k x_{j+1}^{-1} a_k^{-1}$	$1 \le j < k-1 < n-1$
$i > k$	$x_{j+1} x_{k+1} x_{j+1}^{-1} x_{k+1}^{-1}$	$1 \le j < k-1 < n-2$

Table 2

Thus we have

$$C_{n-1} = \langle x_2, \ldots, x_{n-1}, a_1, \ldots, a_{n-1} \mid R, S, T \rangle \quad , \tag{8}$$

where

$$S = \{ x_j x_{j+1} x_j = x_{j+1} x_j x_{j+1} \mid 2 \le j \le n-2 \} \quad , \tag{9}$$

$$T = \{ x_j x_k = x_k x_j \mid 2 \le j < k-1 < n-1 \} \quad ,$$

and R consists of the relations (see Exercise 4)

$$x_k^{-1} a_j x_k = \begin{cases} a_k & , \quad \text{if } j = k-1 \ , \\ a_k a_{k-1} a_k^{-1} & , \quad \text{if } j = k \ , \\ a_j & , \quad \text{otherwise.} \end{cases} \tag{10}$$

278

Now let A_{n-1} and B_{n-1} be the subgroups of C_{n-1} generated by the a's and x's respectively. Then it follows from the results of §20 that

(i) B_{n-1} is correctly named, by (9),

(ii) A_{n-1} is free of rank n-1 (see also Exercise 5),

(iii) C_{n-1} is a split extension of B_{n-1} by A_{n-1} (Exercise 17.3).

A simple induction now gives the following structure theorem.

Theorem 1.   *The algebraical braid group has two chains of sub-groups*

$$E = A_n \leq A_{n-1} \leq \ldots \leq A_2 \leq A_1 = C_1 \leq C_2 \leq \ldots \leq C_{n-1} \leq C_n = B_n \quad ,$$

both terminating in the unpermuted braid group U_n *and such that*

(i)   for $1 < i \leq n$, $A_i \lhd A_{i-1}$ *and* A_{i-1}/A_i *is free of rank i-1 , and*

(ii)   for $1 \leq i \leq n$, $A_i \lhd C_i$ *and* $C_i/A_i \cong B_i$.

Observe that the elements $x_i,\ldots,x_{n-1}$ generate a complement for A_{n-i+1} in C_{n-i+1} $(1 \leq i \leq n)$ and that

$$x_i^2, x_i x_{i+1}^2 x_i^{-1}, \ldots, x_i \ldots x_{n-2} x_{n-1}^2 x_{n-2}^{-1} \ldots x_i^{-1} \tag{11}$$

freely generate a complement $(D_i$, say) for A_{n-i+1} in A_{n-i} $(1 \leq i \leq n-1)$. Letting T be a transversal for U_n in B_n (with elements in one-to-one correspondence with those of S_n), we see that each element of B_n is uniquely expressible in the form $d_1\ldots d_{n-1}t$, with $t \in T$, $d_i \in D_i$ $(1 \leq i \leq n-1)$. This is the first assertion of the next result; the second follows from the first by an independent induction.

Theorem 2.   *The elements of* B_n *admit a normal form, and* B_n *has soluble word problem.*

We now turn our attention to the problem of identifying those elements of U_n that centralize A_{n-1} . Write such an element c in normal form: for some i with $0 \leq i \leq n-1$,

$$c = d_{i+1} \cdots d_{n-1}, \quad d_j \in D_j, \quad d_{i+1} \neq e \quad .$$

Assuming that $c \neq e$, that is, $i \neq n-1$, we claim that $i = 0$.
Suppose for a contradiction that $0 < i < n-1$, and recall the
definitions of u_i and a_i given above. Since $i \neq 0$, it
follows that c commutes with $a_i = x_1 \cdots x_{i-1} x_i^2 x_{i-1}^{-1} \cdots x_1^{-1}$.
But so do all d_j for $j > i+1$ (see (11)), whence

$$d_{i+1} a_i d_{i+1}^{-1} = a_i \quad .$$

Conjugating this equation by u_i^{-1} , it follows that $u_i d_{i+1} u_i^{-1}$
commutes with $u_i a_i u_i^{-1} = a_1$ (see Exercise 6). But since
$u_i d_{i+1} u_i^{-1} \in A_{n-1}$ and A_{n-1} is a free group it follows from
Exercise 2.16 that $u_i d_{i+1} u_i^{-1}$ and a_1 are powers of a common
element. However, a_1 is a free generator and thus not a proper
power, and so there is an $m \in Z$ such that

$$u_i d_{i+1} u_i^{-1} = a_1^m \quad .$$

Since $i \neq 0$, conjugation of this equation by x_1 shows that

$$a_1^m = (x_1^{-1} u_i) d_{i+1} (x_1^{-1} u_i)^{-1} \in D_2 \quad .$$

But $D_2 \cap A_{n-1} = E$, whence $m = 0$ and $d_{i+1} = e$, a contradiction.

Now suppose that in the action of B_{n-1} on A_{n-1} given by
(10), some $b \neq e$ acted as the identity. Then so would its
image in $B_{n-1}/U_{n-1} = S_{n-1}$, which acts simply by permuting the
a_i . Hence, $b \in U_{n-1} = D_2 \cdots D_{n-1}$, in contradiction to the claim
just established. Thus, B_{n-1} acts faithfully on A_{n-1} , that
is, B_{n-1} embeds in Aut A_{n-1} . In view of Exercises 3 and 4, we
have the following result.

Theorem 3. *For all* $n \in N$, $B_n \cong G_n$.

Let $c = d_1 \ldots d_{n-1} \in U_n$ centralize A_{n-1}, and assume that $c \neq e$, that is, $d_1 \neq e$. Since every element of $B_{n-1} \supseteq U_{n-1} = D_2 \ldots D_{n-1}$ fixes the product $a_{n-1} \ldots a_1$, so does d_1, and so (by Exercise 2.16 again) there is an $m \in Z$ such that

$$d_1 = (a_{n-1} \ldots a_1)^m \quad .$$

Hence $c = (a_{n-1} \ldots a_1)^m u_m$, where $u_m = d_2 \ldots d_{n-1} \in U_{n-1}$ is uniquely determined by m (by the above claim) and commutes with $a_{n-1} \ldots a_1$. That such a u_m exists for some $m > 0$ follows from the fact that $Z(B_n) \cap U_n \neq E$ (Exercise 10). We deduce that

$$U_n \supseteq C_{U_n} A_{n-1}) \supseteq Z(B_n) \cap U_n \supseteq <(x_1 \ldots x_{n-1})^n> \quad ,$$

so that our last theorem follows at once from Exercises 9 and 10.

Theorem 4.   *For* $n \geq 3$,

$$Z(B_n) = <(x_1 \ldots x_{n-1})^n> \quad .$$

EXERCISE 1.   Find a regular Nielsen transformation from the basis $(a_1, \ldots, a_n)$ of F_n to $(a_1, \ldots, a_{k+1}, a_{k+1} a_k a_{k+1}^{-1}, \ldots, a_n)$.

EXERCISE 2.   Use the presentation of S_n given in §5 to show that S_n is a homomorphic image of B_n .

EXERCISE 3.   Show that in Aut F_n, $\xi_i \xi_{i+1} \xi_i = \xi_{i+1} \xi_i \xi_{i+1}$, $1 \leq i \leq n-1$, and deduce that C_n is a homomorphic image of B_n .

EXERCISE 4.   Compare the action (10) with that of the generators of $G_{n-1} \leq$ Aut F_{n-1} (see Fig.4).

EXERCISE 5.   Let F be free on $b_1, \ldots, b_{n-1}$ and let x_k

$(2 \le k \le n-1)$ be the automorphism of F given by

$$x_k : (b_1,\ldots,b_{k-1},b_k,\ldots,b_{n-1}) \mapsto (b_1,\ldots,b_k,b_k b_{k-1} b_k^{-1},\ldots,b_{n-1}) \, .$$

If a_k denotes conjugation by b_k $(w \mapsto b_k^{-1} w b_k, \; 1 \le k \le n-1)$, show that the x's and a's satisfy the relations (9) and (10), so there is a homomorphism from the group C_{n-1} of (8) into Aut F . Noting that the image of the subgroup A_{n-1} is just Inn F , deduce that A_{n-1} is free on $a_1,\ldots,a_{n-1}$.

EXERCISE 6.   Use the braid relations to prove that, for $k \in Z$ and $1 \le i \le n-1$, the equation

$$x_1 \ldots x_i x_1 \ldots x_{i-1} x_i^k x_{i-1}^{-1} \ldots x_1^{-1} x_i^{-1} \ldots x_1^{-1} = x_1^k$$

holds in B_n .

EXERCISE 7.   Use the normal form of Theorem 2 to prove that U_n is torsion free.

EXERCISE 8.   Prove that for $n \ge 3$, $Z(B_n) \le U_n$.

EXERCISE 9.   Starting from the trivial braid on n strings, consider the result of rotating the lower frame through an angle 2π in its own plane. Express the resulting braid in terms of $x_1,\ldots,x_{n-1}$ and evaluate its significance.

EXERCISE 10.   Prove that for $n \ge 2$, $z = (x_1 \ldots x_{n-1})^n \in Z(B_n) \cap U_n \backslash E$, and that z is not a proper power in U_n . [Hint: Show by induction that z has normal form

$$(x_1 \ldots x_{n-1} x_{n-1} \ldots x_1)(x_2 \ldots x_{n-1} x_{n-1} \ldots x_2) \ldots (x_{n-1}^2) \quad ,$$

and use the fact that $x_1 \ldots x_{n-1} x_{n-1} \ldots x_1 = a_{n-1} \ldots a_1$ is centralized by $x_2,\ldots,x_{n-1}$].

EXERCISE 11.   Prove that the upper central series of B_n terminates at $Z(B_n)$, that is, $Z(B_n/Z(B_n))$ is trivial.

EXERCISE 12.   Prove that B_3 is isomorphic to the group of the
trefoil knot, and show that $Z(<x,y\,|\,x^3 = y^2>) = <x^3>$.

EXERCISE 13.   Prove that $U_3 \cong F_2 \times F_1$, a direct product of
free groups.

EXERCISE 14.   Show that U_n^{ab} is free abelian of rank $\frac{1}{2}n(n-1)$.

§30. Tangles

A *tangle* is a piece of a knot diagram from which there emerge
just four arcs, pointing in the compass directions NE,NW,SE,SW,
and thus may be constructed by taking a planar graph Γ with
vertex set $V_1 \cup V_2$ such that the members of V_1 are 4-valent
and V_2 consists of four univalent vertices that comprise the
vertices of a square S in R^2 with $\Gamma \backslash V_2 \subseteq S^\circ$, and replacing
each vertex in V_1 by a crossing. Two tangles are called
equivalent if one can be transformed into the other by a finite
sequence of elementary knot deformations (see Fig.28.2). Thus,
for example, all the tangles in Fig.1 lie in different equivalence
classes except for the last two (Exercise 1).

Note that the dihedral group D_4 acts on the set of tangle-
classes in an obvious way, as does reflection in the plane of the

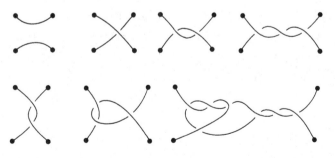

Fig.1

283

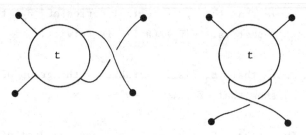

Fig.2

paper (Exercise 2), and that all these operations preserve the
number of crossings. However, we shall be chiefly interested in
the operations illustrated in Fig.2, whose effects on a tangle t
are denoted by ta and tb respectively. It is easy to verify
by inspection (Exercise 3) that if we let O denote the first
tangle in Fig.1, then the other six can be obtained from it by
applying the operations

$$a, a^2, a^3, ab, aba, a^3 b^{-1} a^3 \quad , \tag{1}$$

respectively. Note that the tangles $\{Oab^n \mid n \in Z\}$ are just the
braids on two strings, and that reflection of these in the NW-SE
diagonal yields the *integral* tangles $\{Oa^{n+1} \mid n \in Z\}$. The torus
knots and links of Example 28.5 are obtained from the latter by
joining their top two and bottom two vertices. The integral
tangles from a subset of the *rational tangles*, which are obtained
from O by applying a finite sequence of operations of the form
$a^{\pm 1}, b^{\pm 1}$.

The free group $F = \langle a, b \mid \rangle$ thus acts on the tangle classes,
and in particular on the rational ones, which comprise the orbit
of O . Now any element of F can be written in the form

$$w = b^{j_0} a^{i_1} b^{j_1} \ldots a^{i_{n-1}} b^{j_{n-1}} a^{i_n} \quad , \tag{2}$$

where

$$i_k, j_{k-1} \in Z , \quad 1 \le k \le n \in N \quad , \tag{3}$$

284

and this expression is unique if we specify that

$$i_k \neq 0 \neq j_k \; , \quad 1 \leq k \leq n-1 \; . \tag{4}$$

With any such element, we can associate the continued fraction

$$w^* = i_n + \cfrac{1}{j_{n-1}+} \; \cfrac{1}{i_{n-1}+} \; \cdots \; \cfrac{1}{j_1+} \; \cfrac{1}{i_1} \in Q^* \quad , \tag{5}$$

where $Q^* = Q \cup \{\infty\}$, and ∞ obeys the usual algebraic rules.

Given any tangle t , it is easy to see that $tab^{-1}a$, $tb^{-1}ab^{-1}$ and $tba^{-1}b$ belong to the same class, and so we define the *tangle group* by

$$T = \langle a,b \mid ab^{-1}a = b^{-1}ab^{-1} = ba^{-1}b \rangle \; . \tag{6}$$

Now it is clear that the stabilizer in T of O contains the subgroup $B = \langle b \rangle$. The converse inclusion is also true, and modulo this intuitively reasonable result, we shall prove the rather striking fact that two rational tangles Ow_1, Ow_2 are equivalent if and only if $w_1^* = w_2^*$ in Q^* .

Theorem 1.   *With the above notation, the rule* $w \mapsto w^*$ *defines a one-to-one correspondence between the right cosets of* B *in* T *and the elements of* Q^* .

The proof begins with an examination of the group $SL(2,Z)$, whence we shall approach the theorem by a sequence of lemmas. The *modular group* M is defined to be the quotient of $SL(2,Z)$ by its centre, which just consists of $\pm I_2$ (Exercise 5).

Lemma 1. $M \cong Z_3 * Z_2$.

Proof.   The first step is to show that the matrices

$$b = \begin{pmatrix} 1 & 1 \\ 0 & 1 \end{pmatrix} , \quad y = \begin{pmatrix} 0 & -1 \\ 1 & 0 \end{pmatrix} \tag{7}$$

generate $SL(2,Z)$, and this is done by proving that

$$u = \begin{pmatrix} \alpha & \beta \\ \gamma & \delta \end{pmatrix} \in \langle b,y \rangle, \quad u \in SL(2,Z)$$

by induction on $|\delta|$. When $\delta = 0$, $\beta\gamma = -1$ and so $\beta = \pm 1$.
If $\beta = -1$, then

$$u = \begin{pmatrix} \alpha & -1 \\ 1 & 0 \end{pmatrix} = \begin{pmatrix} 1 & \alpha \\ 0 & 1 \end{pmatrix}\begin{pmatrix} 0 & -1 \\ 1 & 0 \end{pmatrix} = b^{\alpha}y \quad,$$

and if $\beta = 1$, then

$$y^2u = \begin{pmatrix} -\alpha & -1 \\ 1 & 0 \end{pmatrix} = b^{-\alpha}y \quad,$$

so that $u \in \langle b,y \rangle$ in either case. Now let $\delta \neq 0$. If
$|\beta| < |\delta|$, then

$$yu = \begin{pmatrix} -\gamma & -\delta \\ \alpha & \beta \end{pmatrix}$$

belongs to $\langle b,y \rangle$ by induction. Otherwise, there is an $n \in Z$
such that $|\beta+n\delta| < |\delta|$, whereupon

$$yb^nu = \begin{pmatrix} 0 & -1 \\ 1 & 0 \end{pmatrix}\begin{pmatrix} 1 & n \\ 0 & 1 \end{pmatrix}\begin{pmatrix} \alpha & \beta \\ \gamma & \delta \end{pmatrix} = \begin{pmatrix} -\gamma & -\delta \\ \alpha+n\gamma & \beta+n\delta \end{pmatrix}$$

belongs to $\langle b,y \rangle$ for the same reason. It follows that
$\langle b,y \rangle = SL(2,Z)$.

Now define

$$x = by = \begin{pmatrix} 1 & 1 \\ 0 & 1 \end{pmatrix}\begin{pmatrix} 0 & -1 \\ 1 & 0 \end{pmatrix} = \begin{pmatrix} 1 & -1 \\ 1 & 0 \end{pmatrix} \quad, \tag{8}$$

and observe that $x^3 = -I_2$. Thus, M is generated by the elements $\pm x, \pm y$ of orders $3, 2$ respectively. To prove that the
resulting homomorphism $Z_3 * Z_2 \twoheadrightarrow M$ is an isomorphism, we have
to show that no non-trivial word $w = \ldots x^{\pm 1}yx^{\pm 1}y \ldots$ in $SL(2,Z)$
can be equal to $\pm I_2$. By taking inverses, changing sign and
conjugating (by $x^{\pm 1}$ or y), we can assume that a minimal counter-
example begins with $x^{\pm 1}$ and ends with y . Now such a word is

a non-empty product of positive powers of

$$xy = by^2 = -b = \begin{pmatrix} -1 & -1 \\ 0 & -1 \end{pmatrix} \quad , \text{ and}$$

$$x^{-1}y = \begin{pmatrix} 0 & 1 \\ -1 & 1 \end{pmatrix}\begin{pmatrix} 0 & -1 \\ 1 & 0 \end{pmatrix} = \begin{pmatrix} 1 & 0 \\ 1 & 1 \end{pmatrix} = a \text{ , say.} \qquad (9)$$

Thus, $\pm I_2$ is written as a product of elements of the form

$$a^\alpha = \begin{pmatrix} 1 & 0 \\ \alpha & 1 \end{pmatrix} \quad , \quad b^\beta = \begin{pmatrix} 1 & \beta \\ 0 & 1 \end{pmatrix} \quad ,$$

$\alpha, \beta \in \mathbb{N}$. Since it is clear that both types must actually occur, there results a matrix with non-negative entries and trace > 2 (Exercise 6). Since $\text{tr}(\pm I_2) = \pm 2$, this is the desired contradiction.

With any matrix

$$\begin{pmatrix} \alpha & \beta \\ \gamma & \delta \end{pmatrix} \in SL(2,\mathbb{Z}) \quad ,$$

we can associate a *Möbius transformation*

$$z \mapsto \frac{\alpha z + \gamma}{\beta z + \delta} \qquad (10)$$

of the extended complex plane $C^* = C \cup \{\infty\}$. Since it is easy to prove (Exercise 7) that this correspondence is a homomorphism with kernel $\pm I_2$, we let M stand for the group of transformations of the form (10) $(\alpha, \beta, \gamma, \delta \in \mathbb{Z}$, $\alpha\delta - \beta\gamma = 1)$.

Lemma 2. *Under the action of* M *, the stabilizer of* $0 \in C^*$ *is the infinite cyclic group generated by* $b = \pm \begin{pmatrix} 1 & 1 \\ 0 & 1 \end{pmatrix}$ *, and the orbit containing* 0 *is* Q^* *.*

Proof. Under the action of $\pm \begin{pmatrix} \alpha & \beta \\ \gamma & \delta \end{pmatrix}$, 0 is mapped to $(\alpha 0 + \gamma)/(\beta 0 + \delta) = \gamma/\delta \in Q^*$, so the stabilizer of 0 is as claimed, and Q^* contains its orbit. Conversely, a typical member of Q^* has the form γ/δ with $(\gamma, \delta) = 1$, including the possibilities $\{\gamma, \delta\} = \{0, 1\}$. Thus, there exist $\alpha, \beta \in \mathbb{Z}$ with $\alpha\delta - \beta\gamma = 1$, and 0 is mapped to γ/δ by $\pm \begin{pmatrix} \alpha & \beta \\ \gamma & \delta \end{pmatrix} \in M$.

Now regard the matrices a,b given by (9),(7) as members of M , so that by san again, the word w of (2), subject to (3) and (4), may be thought of as a member of M . With this notation and that of (5), we have the following lemma.

Lemma 3. $Ow = w^*$.

Proof.   Any $z \in C^*$ is mapped by a^k, b^ℓ $(k, \ell \in Z)$ to

$$z + k , \quad z/(\ell z + 1)$$

respectively. We proceed by induction on n , noting that the case $n = 1$ is obvious. Now let $n > 1$ and assume the result for $n - 1$, so that if

$$w' = b^{j_o} a^{i_1} b^{j_1} \ldots a^{i_{n-1}} ,$$

then

$$Ow' = i_{n-1} + \cfrac{1}{j_{n-2} +} \cdots \cfrac{1}{j_1 +} \cfrac{1}{i_1} = q ,$$

say. Hence,

$$Ow = Ow' \, b^{j_{n-1}} a^{i_n} = q b^{j_{n-1}} a^{i_n}$$

$$= (q/(j_{n-1} q + 1)) \, a^{i_n}$$

$$= i_n + \cfrac{1}{j_{n-1} +} \cfrac{1}{q} = w^* ,$$

as required.

Theorem 1 will be proved if we can show that M is generated by a,b and that the right-hand side of (6) is a presentation for it. This is done using Tietze transformations in the proof of the final lemma.

Lemma 4. $M \cong T$.

Proof. From the proof of Lemma 1, $M = \langle x,y \mid x^3, y^2 \rangle$, and y is eliminated by the Tietze transformation given by $y = xa$ (9). Thus we obtain $\langle a,x \mid x^3, (xa)^2 \rangle$. Using these two relations in conjunction with (8) and (9), we have

$$b = xy^{-1} = xa^{-1}x^{-1} = x^2a = x^{-1}a .$$

A further Tietze transformation yields

$$\langle a,b \mid (ab^{-1})^3, (ab^{-1}a)^2 \rangle ,$$

which is obviously equivalent to (6).

We leave the reader to draw his own conclusions as to how significant are the satisfying links thus forged between geometry, arithmetic and algebra under the humble auspices of a single group.

EXERCISE 1. Use elementary knot deformations to show that the last two tangles of Fig.1 are equivalent. Check that the continued fractions $1 + \dfrac{1}{1+} \dfrac{1}{1}$ and $3 + \dfrac{1}{-1+} \dfrac{1}{3}$ represent the same member of $Q^* = Q \cup \{\infty\}$.

EXERCISE 2. Show that the group $\Delta(4,2,2) \cong D_4 \times Z_2$ acts in a natural way on the tangle classes.

EXERCISE 3. Check that the tangles of Fig.1 are as described in (1).

EXERCISE 4. Draw the pictures to show that the effect of each of the operations $ab^{-1}a$, $b^{-1}ab^{-1}$ and $ba^{-1}b$ on any tangle is to rotate it through π about the NW–SE diagonal.

EXERCISE 5. Prove that $Z(SL(2,Z)) = \{\pm I_2\}$.

EXERCISE 6.   Let $c \in SL(2,Z)$ be a product of matrices of the
form $\begin{pmatrix} 1 & 0 \\ \alpha & 1 \end{pmatrix}, \begin{pmatrix} 1 & \beta \\ 0 & 1 \end{pmatrix}, \alpha, \beta \in N$, involving each factor at least
once. Use induction on the number of factors to prove that
tr c > 2 .

EXERCISE 7.   Show that the mapping sending $\begin{pmatrix} \alpha & \beta \\ \gamma & \delta \end{pmatrix} \in SL(2,Z)$ to
the Möbius transformation $\frac{\alpha z + \gamma}{\beta z + \delta}$ (acting on C^* on the right) is
an epimorphism with kernel $\pm I_2$.

Guide to the literature and references

Comments are given below on the material in each chapter in turn, and these embrace sources, alternative approaches and suggestions for further reading. The six books referred to by authors' initials are of general interest. That the lists of references are fairly minimal may be excused in view of the extensive and impressive bibliography in [LS].

[CM]   H.S.M. Coxeter and W.O.J. Moser, *Generators and relations for discrete groups*, 4th edition, Springer-Verlag, Berlin-Heidelberg-New York, 1979.

[J]   D.L. Johnson, *Presentations of groups*, Cambridge University Press, 1976.

[LS]   R.C. Lyndon and P.E. Schupp, *Combinatorial group theory*, Springer-Verlag, Berlin-Heidelberg-New York, 1977.

[M]   I.D. Macdonald, *The theory of groups*, Oxford University Press, 1968.

[MKS] W. Magnus, A. Karrass and D. Solitar, *Combinatorial group theory*, Interscience, New York, 1966.

[R]   J.J. Rotman, *The theory of groups: an introduction*, 2nd edition, Allyn and Bacon, Boston, 1973.

1. For the novice, the best introduction to group presentations is contained in Chapter 8 of [M]. The definition of free groups (via the adjoint of the forgetful functor) in Chapter 11 of [R] closely parallels our own, while interesting alternatives are given in Chapter 1 of [MKS] and Chapter 3 of [10]. Permutation groups are used to give a slick proof of the associative law for

free groups in [12], and the method extends to the corresponding (harder) problem for free products.

The Nielsen-Schreier theorem is proved in [7] for finitely-generated subgroups, and in [9] for arbitrary subgroups. That Nielsen's method can be adapted to prove the general result is shown in [2]. Schreier's method is extended in [13] to prove the Kurosh theorem [5] for subgroups of free products. Both these theorems, as well as the (even harder) Grushko-Neumann Theorem [3,6] now boast a number of essentially topological proofs [R] (see also [8]), [1], [4], [11].

[1]   D.E. Cohen, *Combinatorial group theory: a topological approach*, Queen Mary College Mathematics Notes, London, 1978.

[2]   H. Federer and B. Jónsson, Some properties of free groups, *Trans.Amer.Math.Soc.* 68 (1950), 1-27.

[3]   I.A. Grushko, Über die Basen eines freien Produktes von Gruppen, *Mat. Sbornik*, N.S. 8 (1940), 169-182.

[4]   P.J. Higgins, *Notes on categories and groupoids,* Van Nostrand-Reinhold, New York, 1971.

[5]   A.G. Kurosh, Die Untergruppen der freien Produkte von beliebigen Gruppen, *Math.Ann.* 109 (1934), 647-660.

[6]   B.H. Neumann, On the number of generators of a free product, *J. London Math.Soc.* 18 (1943), 12-20.

[7]   J. Nielsen, Om Regning med ikke kommutative Faktoren og dens Anvendeise i Gruppeteorien, *Mat. Tidssk.* B (1921), 77-94.

[8]   J.J. Rotman, Covering complexes with applications to algebra, *Rocky Mountain J. of Math.* 3 (1973), 641-674.

[9]   O. Schreier, Die Untergruppen der Freien Gruppen, *Abh.Math.Sem.Univ. Hamburg* 5 (1927), 161-183.

[10]   J.R. Stallings, *Group theory and three-dimensional manifolds,* Yale University Press, New Haven, 1971.

[11]   J.R. Stallings, A topological proof of Grushko's

theorem on free products, *Math.Zeit.* 90 (1965), 1-8.

[12] B.L. van der Waerden, Free products of groups, *Amer.J.Math.* 70 (1948), 527-528.

[13] A.J. Weir, The Reidemeister-Schreier and Kurosh subgroup theorems, *Mathematica* 3 (1956), 47-55.

2. Presentations for the dihedral and quaternionic (or dicyclic) groups are given in Chapter VII of [1], which also contains an alternative presentation of S_n and a related one for A_n . Another proof of the Basis theorem appears in [M] and [R], while that given here (including the Invariant Factor theorem for matrices) appears in more general form in §16 of [2]. The last theorem in §6 is Theorem 8.16 of [M], where it is proved more or less from scratch.

[1]   R.D. Carmichael, *An introduction to the theory of groups of finite order*, Dover, New York, 1956.

[2]   C.W. Curtis and I. Reiner, *Representation theory of finite groups and associative algebras*, Interscience, New York, 1962.

3. The definition and basic properties of the multiplicator are to be found in [14], and some finite groups with trivial multiplicator and non-zero deficiency appear in [15]. The deficiency problem for metacyclic groups is solved in [16] and [1], and the spectral sequence argument in the latter is obviated by the use of central stem extensions in [2]. The groups of Mennicke, Macdonald and Wamsley appear in [12], [11] and [17], respectively, and the first of these has recently been embedded in a much larger (5-ply infinite) class by Post [13]. The $J(a,b,c)$ are studied in [8], which is based on [18]. $F(2,8)$ and $F(2,10)$ are proved to be infinite in [3], and the bound for $|F(2,9)|$ is given in [5]. The Fibonacci groups are introduced in [9] and studied further in [6], [7] and

several papers by Campbell and Robertson (see the references in [10]). These authors have also found two-
generator, two-relation presentations for the SL(2,p) and
SL(2,8) [4], and the latter is the only known example of
an interesting simple group. A more comprehensive list
of references is given in the survey article [10].

[1]   F.R. Beyl, The Schur multiplicator of metacyclic
        groups, *Proc.Amer.Math.Soc.* 40 (1973), 413–418.

[2]   F.R. Beyl and M.R. Jones, Addendum to 'the Schur
        multiplicator of metacyclic groups', *Proc.Amer.
        Math.Soc.* 43 (1974), 251–252.

[3]   A.M. Brunner, The determination of Fibonacci
        groups, *Bull.Austral.Math.Soc.* 11 (1974), 11–14.

[4]   C.M. Campbell and E.F. Robertson, Two-generator
        two-relation presentations for special linear
        groups; to appear in *Bull. London Math.Soc.*

[5]   G. Havas, J.S. Richardson and L.S. Sterling, The
        last of the Fibonacci groups; to appear in *Proc.
        Royal Soc. Edinburgh.*

[6]   D.L. Johnson, Extensions of Fibonacci groups,
        Bull. London Math.Soc. 7 (1974), 101–104.

[7]   D.L. Johnson, Some infinite Fibonacci groups,
        Proc. Edinburgh Math.Soc. 19 (1975), 311–314.

[8]   D.L. Johnson, A new class of 3-generator finite
        groups of deficiency zero, *J. London Math.Soc.* 19
        (1979), 59–61.

[9]   D.L. Johnson, J.W. Wamsley and D. Wright, The
        Fibonacci groups, *Proc. London Math.Soc.* 29 (1974),
        577–592.

[10]   D.L. Johnson and E.F. Robertson, Finite groups of
        deficiency zero, in *Homological group theory* (ed.
        C.T.C. Wall), Cambridge University Press, 1979.

[11]   I.D. Macdonald, On a class of finitely-presented
        groups, *Canad.J.Math.* 14 (1962), 602–613.

[12]   J. Mennicke, Einige endliche Gruppen mit drei

Erzengenden und drei Relationen, *Arch.Math.* 10
(1959), 409-418.

[13] M.J. Post, Finite three-generator groups with zero
deficiency, *Comm. in Alg.* 6 (1978), 1289-1296.

[14] I. Schur, Untersuchungen über die Darstellung der
endlichen Gruppen durch gebrochene lineare Substi-
tutionen, *J. Reine Angew.Math.* 132 (1907), 85-137.

[15] R.G. Swan, Minimal resolutions for finite groups,
Topology 4 (1965), 193-208.

[16] J.W. Wamsley, The deficiency of metacyclic groups,
Proc.Amer.Math.Soc. 24 (1970), 724-726.

[17] J.W. Wamsley, A class of three-generator three-
relation finite groups, *Canad.J.Math.* 22 (1970),
36-40.

[18] J.W. Wamsley, Some finite groups with zero deficiency, *J.Austral.Math.Soc.* 18 (1974), 73-75.

4. The coset enumeration process made its debut in [1],
and a description of the modified method of §13 appears
in the fourth edition of [CM].

[1] J.A. Todd and H.S.M. Coxeter, A practical method
for enumerating cosets in an abstract finite group,
Proc. Edinburgh Math.Soc. (2) 5 (1936), 25-36.

5. The triangle groups were first studied intensively
in [1], and a more up-to-date treatment is to be found
in [CM]. A purely algebraic proof that $D(\ell,m,n)$ is infinite if and only if $1/\ell + 1/m + 1/n \leq 1$ is given
in [4]. A much more general (and possibly more natural)
class is that of co-compact Fuchsian groups, treated in
detail in [2] and [3].

[1] R. Fricke and F. Klein, *Vorlesungen über die
Theorie der automorphen Funktionen*, Vol.I, Teubner,
Leipzig, 1897.

[2] A.M. Macbeath, *Fuchsian groups*, Queen's College
Notes, Dundee, 1961.

[3]   W. Magnus, *Noneuclidean tesselations and their groups*, Academic Press, New York-London, 1974.

[4]   G.A. Miller, Groups defined by the orders of two generators and the order of their product, *Amer.J. Math.* 24 (1902), 96-100.

6.   The account of extension theory given here is adapted from [3], whose sequel [4] contains a study of the case when the kernel of the extension is non-abelian.   The connection with cohomology is also covered in [7] and, from a new and revealing point of view in [6].   Homological matters are treated exhaustively in [1], and their relevance to group theory and algebraic number theory is exploited in [6] and [2], respectively.   The proof of the Golod-Šafarevič theorem [5] is due to P. Roquette [2,6].   Kostrikhin's example is mentioned in [7], which also contains an excellent account of the theory of the multiplicator, including the result on direct products. The minimal presentation for a wreath product is to be found in [8], and a more general result in [9], which gives a general survey of minimal presentations.

[1]   H. Cartan and S. Eilenberg, *Homological algebra*, Princeton University Press, 1956.

[2]   J.W.S. Cassels and A. Fröhlich (eds.), *Algebraic number theory*, Academic Press, London, 1967.

[3]   S. Eilenberg and S. Maclane, Cohomology theory in abstract groups I, *Ann. of Math.* 48 (1947), 51-78.

[4]   S. Eilenberg and S. Maclane, Cohomology theory in abstract groups II, *Ann. of Math.* 48 (1947), 326-341.

[5]   E.S. Golod and I.R. Šafarevič, On the class field tower, *Izv.Akad.Nauk SSSR*, 28 (1964), 261-272.

[6]   K.W. Gruenberg, *Cohomological topics in group theory*, Springer, Berlin, 1970.

[7]   B. Huppert, *Endlich Gruppen I*, Springer, Berlin, 1967.

[8]   D.L. Johnson, Minimal relations for certain wreath
       products of groups, *Canad.J.Math.* 22 (1970), 1005-
       1009.

[9]   J.W. Wamsley, Minimal presentations for finite
       groups, *Bull. London Math.Soc.* 5 (1973), 129-144.

7. The material in §23 is derived from [8], and the
fundamental ideas described in §24 appear in much the
same form in the epoch-making article [4], where the
author gives credit to the pioneering work of M. Dehn,
and to the more recent results obtained by M. Greendlinger
and others on groups satisfying the 'metric' small cancellation conditions. The material in §§25,26 has been
pinched more or less wholesale from [3],[2], respectively,
and serve as samples of the numerous applications of the
theory. The existence of free subgroups is discussed in
[7], while §26 can also be found in [1], which includes
many other applications. The notes [5] contain an account
of decision problems in group theory, and W.W. Boone's
(negative) solution of the word problem is discussed in
[R]. Small cancellation theory is applied to the con-
jucacy problem in [6], while [LS] contains a thorough and
up-to-date account of the state of the art.

[1]   C.P. Chalk, *Small cancellation theory and appli-
       cations to the Fibonacci groups*, Ph.D. thesis,
       University of East Anglia, 1976.

[2]   C.P. Chalk and D.L. Johnson, The Fibonacci groups
       II, *Proc. Royal Soc. Edinburgh* 77A (1977), 79-86.

[3]   D.J. Collins, Free subgroups of small cancellation
       groups, *Proc. London Math. Soc.* (3) 26 (1973),
       193-206.

[4]   R.C. Lyndon, On Dehn's algorithm, *Math.Ann.* 166
       (1966), 208-228.

[5]   A.W. Mostowski, *Decision problems in group theory,
       Technical report no.19*, University of Iowa, 1969.

[6]   P.E. Schupp, On the conjugacy problem for certain

quotient groups of free groups, *Math.Ann.* 186 (1970), 123-129.

[7] V.V. Soldatova, The sieve method in group theory, *Mat.Sbornik* N.S. 25 (1949), 3-50.

[8] E.R. van Kampen, On some lemmas in the theory of groups, *Amer.J.Math.* 55 (1933), 268-273.

8. The proof of the classification theorem for surfaces is a paraphrase of that in [11], while a refreshing alternative is to be found in [16]. The latter also contains an approach to topological ideas pitched at fairly intuitive level, while [8] provides a more rigorous introduction. The books [5] and [13] comprise the old and new testaments of knot theory, respectively, and [12] is apocryphal only by analogy. As to survey articles on knots, [6] and [14] are sound introductions, while [7] is a more up-to-date account. Braids were invented by E. Artin [1] and [10] provides an excellent survey; braids and related topics are covered extensively in [2]. The algebraic approach to unravelling the structure of braid groups is based on the article [3]. Tangles were invented by J.H. Conway [4], and subsequent progress is reported in [9]. The poem on page 10 of [15] contains some non-trivial information.

[1] E. Artin, Theorie der Zöpfe, *Abh.Math.Sem.Univ. Hamburg* 4 (1925), 47-72.

[2] J. Birman, Braids, links and mapping class groups, *Ann. of Math. Studies no.82*, Princeton University Press, Princeton, 1974.

[3] W.-L. Chow, On the algebraical braid group, *Ann. of Math.* (2) 49 (1948), 654-658.

[4] J.H. Conway, An enumeration of knots and links, and some of their algebraic properties, in *Computational problems in abstract algebra*, pp.329-358, Pergamon, New York, 1970.

[5] R.H. Crowell and R.H. Fox, *Introduction to knot*

theory (re-issue of 1963 text), Springer, Berlin, 1978.

[6]   R.H. Fox, A quick trip through knot theory, in *Topology of 3-manifolds and related topics*, pp.120-167, Prentice-Hall, Englewood Cliffs, 1962.

[7]   C.McA. Gordon, pp.1-60, in *Knot theory*, Springer Lecture Notes, no.685, Berlin, 1978.

[8]   M. Henle, *A combinatorial introduction to topology*, Freeman, San Francisco, 1979.

[9]   L.H. Kauffman, The Conway polynomial, to appear.

[10] W. Magnus, Braid groups: a survey, in *Proc. Second Internat.Conf. on Theory of Groups (Canberra, 1973)*, pp.463-487, Springer Lecture Notes, no.372, Berlin, 1974.

[11] W. Massey, *Algebraic topology: an introduction*, Harcourt, Brace and World, New York, 1967.

[12] L.P. Neuwirth, Knot groups, *Ann. of Math. Studies* 56, Princeton University Press, Princeton, 1965.

[13] D. Rolfsen, *Knots and links*, Publish or Perish, Berkeley, 1978.

[14] H. Siefert and W. Threlfall, Old and new results on knots, *Canadian J.Math.* 2 (1950), 1-15.

[15] Manifold, University of Warwick, Summer, 1972.

[16] E.C. Zeeman, *The classification theorem for surfaces*, University of Warwick, 1966.

Notes

6. Green's Theorem appeared in: J.A. Green, On the number of automorphisms of a finite group, *Proc.Roy.Soc. London Ser. A* 237 (1956) 574-581.

8. The proof of Theorem 30.1 given here was worked out in collaboration with M. Al-bar.

Index of notation

N	natural numbers
N_o	$N \cup \{0\}$
Z	integers
Q	rationals
R	reals
R^n	n-dimensional Euclidean space
R^n_+	$\{(x_1,\dots,x_n) \in R^n \mid x_1 \geq 0\}$
C	complex numbers
$n!$	n factorial, $n \in N$
$\binom{m}{n}$	binomial coefficient, "m choose n"
(m,n)	highest common factor of $m,n \in N$
$[m/n]$	integer part of fraction m/n
$R[t]$	polynomials over a commutative ring R with identity
$\phi_k(t)$	kth cyclotomic polynomial
$\nabla(z)$	Conway potential function
$A(x)$	Alexander polynomial
$f*g$	resolvent of polynomials f,g
$GF(p)$	Galois field of p elements, p a prime
RG	group ring of G over R
$GL(n,R)$	group of non-singular n×n matrices over ring R
$\det A$	determinant of matrix A
$SL(n,R)$	$\{A \in GL(n,R) \mid \det A = 1\}$
$\leq$	less than or equal to, is a subgroup of
$<$	less than, is a proper subgroup of
$\triangleleft$	is a normal subgroup of
$\dot\cup$	disjoint union of sets
$\setminus$	set-theoretic difference

$\emptyset$	empty set
δ_{ij}	Kronecker delta
©	commutative face of diagram, copyright
$\longrightarrow$	mapping
$\dashrightarrow$	mapping whose existence is alleged
$\mapsto$	effect of mapping on an element
$\blacktriangleright\!\!\blacktriangleright$	injection, monomorphism
$\twoheadrightarrow$	surjection, epimorphism
$\cong$	isomorphism
$\exists!$	there is a unique
1_X	identity mapping on set X
inc	inclusion mapping
nat	natural homomorphism
Ker	kernel of a homomorphism
Im	image of a mapping, homomorphism
$\mathrm{Hom}_G(A,B)$	the group of G-homomorphisms between G-modules A,B
$Z^n(G,A)$	n-dimensional cocycles from group G to G-module A
$B^n(G,A)$	n-dimensional coboundaries
$H^n(G,A)$	nth cohomology group of G with coefficients in A
$A \oplus B$	direct sum of A and B
$\Phi(A)$	intersection of maximal submodules of A
$d(A)$	minimal number of generators of A
P_A	A/pA
A^G	maximal G-trivial factor-module of A
A_G	maximal G-trivial submodule of A
e	identity element of G
E	trivial group
$\ell(w)$	length of reduced word w
$[x]$	order of $x \in G$, modulus of $x \in C$
x^y	conjugate $y^{-1}xy$
$[x,y]$	commutator $x^{-1}y^{-1}xy$

$\lvert X \rvert$	cardinality of set X, underlying polyhedron of simplicial complex X
X^{-1}	$\{x^{-1} \mid x \in X\}$, $X \subseteq G$
$X^{\pm 1}$	$X \cup X^{-1}$
$\langle X \rangle$	subgroup of G generated by $X \subseteq G$
$\bar{R}$	normal closure of $R \subseteq G$
$\langle X \mid R \rangle$	group on generators X with relators R
$F(X)$	free group on X
$A(X)$	free abelian group on X
X^Y	$\{x^y \mid x \in X, y \in Y\}$, for $X, Y \subseteq G$
$[X,Y]$	$\{[x,y] \mid x \in X, y \in Y\}$, for $X, Y \subseteq G$, the subgroup generated by this set, when $X, Y \le G$

G'	derived group of G
G^{ab}	"G abelianized", G/G'
$Z(G)$	centre of G
$\Phi(G)$	Frattini subgroup of G
G^p	$\langle \{x^p \mid x \in G\} \rangle$, p a prime
$M(G)$	Schur multiplicator of G
Aut G	automorphism group of G
Im G	group of inner automorphisms of G
$\gamma_n(G)$	nth term of lower central series of G

$d(G)$	minimal number of generators of G
$r(G)$	$d(G) + d(M(G))$
$r'(G)$	minimal number of relations needed to define G
def G	$\max\{\lvert X \rvert - \lvert R \rvert \mid \langle X \mid R \rangle$ a finite presentation of G$\rangle$
G_p	the class of finite p-groups for which $r(G) = r'(G)$

+	definition	
∔	bonus information	anotations in a coset
∦	known information	enumeration table
∰	coset collapse	

X+	adjunction of a generator	
X−	removal of a generator	
R+	adjunction of a relator	Tietze transformations
R−	removal of a relator	

$|G{:}H|$ index of subgroup $H \leq G$

$G \times H$ direct product of groups, Cartesian product of sets

$G * H$ free product of groups

$G \wr H$ wreath product of groups

$G \otimes H$ tensor product of groups

S_A group of all permutations of a set A

S_∞ S_Z

S_n symmetric group of degree n

Z_n cyclic group of order n

D_n dihedral group of degree n

Q_{2n} quaternionic group of order 4n

A_n alternating group of degree n

Z infinite cyclic group

B_n algebraical braid group on n strings

G_n geometrical braid group

U_n unpermuted braid group

$G_n(w)$ cyclically presented group

$A_n(w)$ $G_n(w)^{ab}$

$F(r,n)$ Fibonacci group

$F(r,n,k)$ generalized Fibonacci group

$E(r,n)$ extended Fibonacci group

$D(\ell,m,n)$ von Dyck group

$\Delta(\ell,m,n)$ triangle group

$Mac(a,b)$ Macdonald group

$M(a,b,c)$ Mennicke group

$W_\pm(a,b,c)$ Wamsley groups

$d(P)$	degree of vertex P
$d(D)$	degree of face D
$M^{\cdot}$	boundary of map M
$\sum^{\cdot}$	sum over boundary vertices or faces
$T(k)$ $\left.\right\}$	small cancellation conditions, $k \in \mathbb{N}$
$C(k)$	

χ	Euler characteristic
∂	topological boundary
#	connected sum of surfaces
$\pi_1(X)$	fundamental group of path-connected space X
$H_1(X)$	$\pi_1(X)^{ab}$
S^n	n-sphere
T_n	connected sum of n tori, $n \in \mathbb{N}_o$
P	real projective plane
P_n	$T_n \# P$
K_n	$T_n \# P \# P$
D	disc
M	Möbius band
C	cylinder

Index

homology group   258,266

homotopy   253

Hopfian group   22

Icosahedron   136

identification code   252,258

identification space   248

infinite dihedral group   197

injective   9

interesting group   63

invariant factors   55

inverse of o-knot   261

inversion in a circle   143

irreducible module   178

isomorphism problem   39

Jacobson radical   180

Jordan curve theorem   213

Klein bottle   247

knot   258

Knott   214

knot group   37,262

knot type   259

Kostrikhin   200

Lexicographic ordering   10

link   268

Listing's knot   259,265

localization   147

locally Euclidean   246

loop   133,215,253

lower central series   201

Lucas numbers   60

Lyndon   238,244

Macdonald   70

Magnus   230

manifold   245

map   218

Mennicke   70

metacyclic group   63,84

Miller   70

minimal resolution   183,186

minor   55

Möbius strip   247

Möbius transformation   287

modular group   114,285

module   147

monoid   256

Moser   86

multiplication table   32

Neighbourhood   245

Nielsen transformation   19,281

Nielsen-Schreier theorem   24,
   60

nilpotency class   73,201

nilpotent group   73,202

nilpotent ideal   180

normal form   279

Obverse of o-knot   261

octahedron   136,231

o-knot   260

one-relator group   118,257

orbit   179,185

orientable surface   247

orthogonal circles   144

Path   253